NOVEL PROCESSES FOR ADVANCED MANUFACTURING

Novel Processes for Advanced Manufacturing

Edited by

Harinirina Randrianarisoa, Ph.D.

ARCLER PRESS

www.arclerpress.com

Novel Processes for Advanced Manufacturing

Harinirina Randrianarisoa, Ph.D.

Arcler Press
2010 Winston Park Drive,
2nd Floor
Oakville, ON L6H 5R7
Canada
www.arclerpress.com
Tel: 001-289-291-7705
001-905-616-2116
Fax: 001-289-291-7601
Email: orders@arclereducation.com

ISBN: 978-1-77361-236-2 (Hardcover)

About the Editor

Harinirina Randrianarisoa, Ph.D.

Hari Randrianarisoa obtained is Ph.D from Institute of Electronics, Microelectronics and Nanotechnology (IEMN) of the University of Science and Technology – Lille (France). His thesis focused for designing and manufacturing of thermal micro sensors. His interests span from on renewable energy especially solar energy. He has a strong knowledge in technology transfer field. He is currently working as a Business advisor for economic development organization in Québec City and as a consultant Scientist advisor.

Contents

List of Figures

List of Tables

Preface

Since the early beginning of industrial revolution, manufacturers are consistently trying to improve the characteristics of their products and efficiency of the manufacturing processes. Man started manufacturing from primitive style to produce very simple products to make his life easier. There were no machines and devices all the manufacturing work was done by hand and the ordinary things of daily use such as axe, hammer and chisels were used as tools for assistance in manufacturing operation. As, time passed, man discovered many new phenomenon of nature, the knowledge of science, technology, engineering and management evolved and systematic studies on manufacturing were started. Today, Manufacturing has grown to a tremendous level, it has become a vast subject. It has become an inter-disciplinary approach which is an integration of many sciences, engineering, management, commerce and finance disciplines etc.

The most recent couple of years have seen unparalleled changes all through the world. Quick changes in the business sectors demand variety of product life cycles and top notch products at low costs. Clients now demand exceptionally high levels of quality and performance. This global trend and market competition has motivated product manufacturing firms to search for dynamic and mechanized manufacturing processes in all fields. In this way many manufacturing processes, technologies and systems have came in existence and research is still continued in this field and many new, radically different manufacturing processes, technologies and systems are expected to come into being in the coming years.

The aim of this book is to provide a description of modern manufacturing practices and novel processes which are speedily replacing the conventional manufacturing processes and systems. This book presents a picture of the globally changing manufacturing world along with the latest market challenges and how they are being addressed.

This book is of interest for graduate level audience who are keen to get in touch with new manufacturing systems, processes and technologies. This book is also of interest for manufacturers and technical staff working in today's innovative manufacturing corporate world. The first chapter provides a general view of advanced manufacturing systems, processes technologies and techniques which are used for controlling them. In the later chapters advances in material forming processes, joining processes, processing techniques, machining, material removal techniques, additive manufacturing, manufacturing process management , manufacturing process control are described and green manufacturing science, engineering, processes, technologies, implementation planning and framework have been described.

Editor

CHAPTER 1

Advanced Manufacturing System and Novel Processes

BACKGROUND AND INTRODUCTION

The global market of the twenty-first century is advancing into one of consolidating national markets, divided customer markets, and quickly changing product innovation technologies. These progressions are driving firms to contend, at the same time, along a few distinct measurements: design, production, supply, sales, marketing and others. Despite the fact that product manufacturing has not been used as a focused weapons truly, the commercial center of the twenty first century forces that product manufacturing accepts an urgent part in the new progressive world.

Advances in human culture have been geared by the production of new materials and innovative processes. The most recent couple of years have seen unparalleled changes all through the world. Quick changes in the business sectors demand definitely abbreviated item life cycles and top notch items at low costs. Clients now lean toward a huge assortment of items. This wonder has motivated product manufacturing firms to search for dynamic and mechanized manufacturing processes in all fields. In this way large scale manufacturing is being supplanted by low-volume, high-assortment manufacturing. Production firms have perceived the significance of adaptability in the assembling framework to address the difficulties postured by the multi-demand market. The idea of flexible manufacturing systems has accomplished noteworthy significance in addressing the difficulties for an assortment of products of shorter lead times, together with higher profitability and quality. The flexibility is the hidden idea driving the move from conventional strategies for manufacturing to the more robotized and integrated processes. They force that organizations executing robotization tasks should set their requirements for various flexibilities under long-term goals. Increasing worldwide competition and quick progression of production innovation are two realties in the present day business condition. These have consolidated to move the business key needs toward quality, cost viability and responsiveness to market changes. (Waldemar Karwowski, n.d.)

The mission of bringing down working expenses and enhanced product efficiency has constrained countless firms to come on advanced manufacturing systems of different sorts. Sensational advancements in advanced manufacturing system at different hierarchical levels can be credited to various advantages that enhance the impressive position of the that organizations. Advanced manufacturing system affect not simply production, but rather all the business operations, giving new challenges to a firm's capacity to manage both production and automation. (Beatty, n.d.)

Advanced manufacturing system can likewise give particular upper hands in cost and process initiative. Events of the most recent decade, for example, the US profitability issues, Japanese production stories of overcoming production problems and the fast paced worldwide economy, have moved manufacturing methodology and process innovation issues from the base to the highest point of the company's need list. The issues encompassing manufacturing advances and their usage have accepted more prominent significance in the production system debates. Specialists and analysts have solid enthusiasm for how advanced manufacturing system can be utilized as a competitive instrument in the worldwide economy. (Bessant, n.d.) A large number of firms are presently using advanced manufacturing system to adapt with latest challenges in the present competitive condition, for example, divided mass markets, shorter item life cycle and increased interest for customization. In spite of the fact that advanced manufacturing system can enable makers to contend under these market conditions, they regularly fill in as a twofold edged sword, forcing organizational difficulties while giving unmistakable upper hand when effectively executed.

Universal business techniques as often as possible demand the exchange of manufacturing technologies. production process is characterized as any setup for making a product, including the general population, hardware, material information sources, methodology and programming in that system. An issue of significance in the key decisions is whether the manufacturing technology should be exchanged without adjustment or should be somewhat modified. Because of the exceptional worldwide competition in manufacturing, producers need to increment their level of competitiveness in the worldwide market. Some production organizations, along these lines, are compelled to experience a time of change so as to contend more viably. Under these conditions, advanced manufacturing system is considered as methods for enhancing competitiveness. The term advanced manufacturing system means computer assisted advances in design, fabrication, transportation and testing, and so on. All in all, advanced manufacturing system can be sorted into two standard ways: The established continuum of essential manufacturing processes which broadens from make-to-arrange production to consistent production; and the level of mix of the general manufacturing setup. Advanced manufacturing system furnishes a firm with a chance to effectively battle commercial market elements and make for itself an upper hand. (Waldemar Karwowski, n.d.) Producers and scholastic experts foresee that advanced manufacturing systems can decrease working bunks, give elevated amounts of yield by expelling human effort, enhance production flexibility and lead time to showcase.

The major areas of concern in implementation of advanced manufacturing technologies for utmost advantage are stated below

Adoption of Advanced manufacturing technologies

(iv) Planning and Implementation of advanced manufacturing systems

(vii) Benefits and Disadvantages of advanced manufacturing system

Advanced manufacturing systems provide new competitive advantages to firms for growth and development. Small firms also benefit from the advanced manufacturing in local context. .

The ability of this innovative system to change production parameters rapidly and precisely implies that organizations can modify their items and accomplish financial aspects of extension in view of low volume and minimal effort production. (Beatty, n.d.)While conventional technology has been seen just as a device in executing business procedure, advanced manufacturing system can possibly straightforwardly influence the company's technology decisions. Advanced manufacturing system offers firms the potential to seek after new inventive methodologies. (Waldemar Karwowski, n.d.)

ADVANCED MANUFACTURING

Advanced manufacturing is the utilization of modern innovative processes to produce high quality products. Advanced Manufacturing centers around enhancing methodologies, procedures and learning that improve profitability. The manufacturing practices can be categorized into five key groups as follows:

- Processes which are applied to alter the shape of the material:

These processes use energy in the form of heat or mechanical force or a combination of these to impart specific shapes to materials.

These processes comprise of

- Casting
- Hot and cold forming processes e.g. forging, rolling, drawing and electroforming, etc.
- Sheet metal working processes e.g. bending and drawing etc.
- Powder forming processes e.g. sintering and plastic molding, etc.
- Processes which are used to machine the products to specific sizes

These processes use tools and technologies to remove material from the parts to make them up to specific sizes an dimensions.

- Conventional machining processes include turning, drilling, reaming, milling and polishing processes etc.
- Non-conventional machining processes include ultrasonic machining, electro chemical machining, electro chemical

grinding, electric discharge machining, laser beam machining and magnetic abrasive finishing, etc.

- Processes which are used for treating surface of the materials

These processes are applied to only the surface of materials to produce specific properties on the surface. These processes include cleaning treatments, surfacing treatments such as shot peening, diffusion and ion transplant etc., coating treatments such as electroplating and galvanizing etc., and advanced thin film depositing treatments e.g. physical and chemical vapor deposition processes etc.

- Processes which are used for connecting parts of materials with each other:

These processes use heat, mechanical fasteners or chemical adhesives for joining the parts with each other to form a complicated assembly. These processes include welding, brazing, sintering, riveting screw joining and adhesive bonding, etc.

- Processes which are used for improving the inherent properties of the materials:

There processes use elevated temperatures, heating and cooling treatments for improving the physical and mechanical characteristics of the materials. These processes include annealing, hardening, and sintering, etc.

At the instant diverse systems of Advanced Manufacturing are present. certain of these systems are well reputed and extensively tested for example

- flexible transfer lines,
- flexible manufacturing cells
- flexible manufacturing systems

Other advanced manufacturing systems are somewhat fresh and they are under studies under various research projects. Many tools and machine manufacturers recommend customized advanced processes for using their tools and machines

DEDICATED MANUFACTURING SYSTEMS

Dedicated Manufacturing Systems are those systems that are considered, outlined and overseen aptly on the necessities of a product or an exceptionally limited group of products. The primary qualities of these systems are:

Rigid Equipment

The equipment is intended to fulfill the requirements of the product, or the limited group of products, to which the entire framework is dedicated. In this manner the machines and gadgets such as transporters are intended to fulfill a exceptionally constrained arrangement of operations that can't be regularly expanded unless extensive expenses are brought about. Stations in exchange lines are a normal case of rigidity of equipment. Typically machine motion is not numerically controlled by a PC but rather mechanically by methods for cams or other mechanical gadgets. (Company, 2004)

High Production Rates

The equipment is dedicated and regularly intended to limit product manufacturing times. Keeping in mind the end goal to cut manufacturing times, at least one operation can be performed in parallel. As a result machines are for the most part fast permitting the system to achieve higher production rates contrasted with other Manufacturing Systems.

Low Skills

The skills required to operate the system are regularly low since human occupations are lessened to stacking and emptying parts, support and repair.

Easy Management

Given the set number of products a Dedicated Manufacturing System produces, and the straightforwardness of work streams in the system, the planning, booking and scheduling of assets is very easy.

Low Investment

The equipment is inflexible and everything is planned to finish just the operations that are important to make the products to which the Dedicated Manufacturing System is dedicated. So, the venture cost of the system is not huge if contrasted with that of flexible systems with computerized machines.

Excess Capacity

The measure of system capacity unused because of decreased demand can't be utilized to make other products. For this reason the remaining value of the investment is little.

There are basically two classifications of Dedicated Manufacturing Systems practically:

Dedicated Machines and Dedicated Flow Lines.

Dedicated machines are those machines pertinently intended to perform proficiently the product manufacturing cycle; these machines are for the most part considered and created in the firm on the grounds that a high learning and experience is required on the manufacturing process. (Company, 2004) Dedicated machines work in stand-alone mode and are totally computerized with the exception of the stacking and un-loading of parts; along these lines they are likewise easy to oversee not requiring any complex devices. **Dedicated manufacturing flow lines** are significant and extensively used sort of Dedicated Manufacturing Systems.

FLEXIBLE MANUFACTURING SYSTEMS

Flexible manufacturing system is a mechanized manufacturing system with an insignificant human activity proficient of creating any part variety resembling with a pre-specified product family; these systems are for the most part used for the production of little or medium volumes, in different sizes of lots that contrast in their structure. The system adaptability is for the most part constrained to the group of part types on which the system is planned and designed. The Flexible manufacturing system has gadgets for scheduling the manufacturing, planning the assets and sparing the production information. As the above definition calls attention to, the primary qualities of Flexible manufacturing systems are:

Flexible equipment

The equipment is sufficiently flexible to fulfill the requirements of the considerable number of products which are members of the family. To be sure all the machines are CNC type and can be modified to perform many manufacturing operations. In most practical instances, it is just required the capability to compose a straightforward PC program to code the manufacturing process cycle of a product into directions that the numerical control of the machine can read, understand and work to act on them. (Company, 2004)

Low Production Rates

Machines have for the most part a shaft for executing manufacturing operations consecutively. As an outcome, machines are for the most part moderate in speed in correlation with machines of Dedicated Manufacturing System. Later developments such as high shaft machines and linear motors are quickly dispersing in Flexible Manufacturing Systems accordingly diminishing handling times and interactive fast motions are made possible.

Medium/High skills

The abilities required to operate the system requires a base information in programming and overseeing CNC machines.

Complex Management

The administration of Flexible manufacturing systems is complex because of the huge number of products. To be sure, for every product it is important to plan appropriate machines, installations and devices.

Large Investment

Machines are flexible and require extensive speculations. Subsequently, the venture cost of the system is huge if contrasted with that of dedicated manufacturing systems.

Excess Capacity

The measure of manufacturing capacity unused due to low demand can be utilized to make diverse products. The remaining value of the investment considers this problem.

ADVANCED MANUFACTURING PROCESS TECHNOLOGIES

Advanced Manufacturing technology is a family of manufacturing technologies which

- Rely upon the utilization and coordination of data, robotization, calculation, and programming, detecting, and networking.
- Make utilization of advanced materials and developing abilities empowered by the physical and organic sciences, for instance nanotechnology. This includes both better approaches to produce existing products, and particularly the production of new products rising up out of new advancements in technology

Advanced Manufacturing is not restricted to rising advancements; rather, it is made out of proficient, productive, profoundly coordinated, firmly controlled processes over a range of internationally focused producers and suppliers. For cutting edge manufacturing to quicken and flourish, it will require the dynamic cooperation of community groups, instructors, specialists, and organizations, as well as Federal, State, and interacting governments. (Beatty, n.d.)

The manufacturing process technologies used for advanced manufacturing comprise of:

- Computer technologies
- High Performance Computing, modeling, simulation and analysis
- Rapid prototyping
- High Precision technologies
- Information technologies
- Advanced robotics and intelligent production systems
- Automation
- Control systems

- Green processes and technologies
- New industrial platform technologies
- Custom manufacturing technologies

BENEFITS OF ADVANCED MANUFACTURING

There exist a number of scientific and competitive aspects that encompass advanced manufacturing systems. Major benefits of adopting advanced manufacturing include

- Less product design and development time
- Labor costs are reduced
- Material costs are reduced
- Organization remains competitive
- Organization may gain tax incentives by adopting environment friendly Advanced manufacturing systems
- Better finance management because of automation of processes
- Robust products and flexible manufacturing setup
- Environment friendly production
- Better control over Safety and health concerns
- improved profitability
- Better plant performance with minimal losses
- customer satisfaction

These benefits pose a wide range, strategic effect on the manufacturing firm and they improve nearly all main areas of a company's business operation.

Advanced manufacturing systems encompass a broad-range of present day computer-based frameworks dedicated to the enhancement of manufacturing operations and consequently up gradation of the firm's business agility. Advanced manufacturing systems, with its many variants, have been credited with the potential to offer, in addition to other things, prior access to market, speedier reactions to changing client needs, and higher quality items with enhanced consistency and unwavering quality, However, consequences of a few practical studies demonstrate that, while most firms accomplish a few advantages, a significant number of them are not completely benefiting their Advanced manufacturing system touted abilities. Since the specialized abilities of Advanced manufacturing systems are well demonstrated, inability to accomplish the potential advantages has been credited to foundation issues, for example, lacking hierarchical arranging and planning for the selection of the Advanced manufacturing systems or sets a life-cycle process which comprises of the accompanying three stages: pre-installation (arranging and legitimization), establishment and startup (observing and assessment). Usage is commonly seen as a mix of the activities in the establishment and startup and post-start up stages.

DISADVANTAGES OF ADVANCED MANUFACTURING

Organizations have distinguished some key issues emerging from actualizing an advanced manufacturing system project and examine the accompanying issues. These issues are

- Extent of work advanced manufacturing system projects can undertake
- Abstraction and demonstration
- Cell design
- Cell operational coordination
- Worker issues in advanced manufacturing systems.

Customarily there has been little consciousness of the need to interface innovation issues with human asset administration issues. Innovations have been used to enhance a firm's project execution capabilities autonomous of works and practices that include laborers in basic management decisions. Firms execute new innovations to enhance their competitive abilities. Decreases in labor costs are made conceivable by new robotized machines and hardware which assume control choices about manufacturing and perform at quick rates operations earlier performed by people. Speedier performing advancements allow increased manufacturing utilizing the same less, or less skilled workers.

The issues associated with implementation of advanced manufacturing systems can be grouped into five sets. The first sets includes general cost- associated problems related with advanced manufacturing system implementation, comprising the cost of principal investment, the cost of manufacturing system procurement, the cost of auxiliary equipment procurement, the cost of developing the operating software, and cost of maintenance and repair. Other four sets are associated with government policies, worker marketplace limitations, in-house company problems and limitations in the marketplace for statistics.

Each of these additionally builds the expenses of installing advanced manufacturing systems-yet the causes are to some degree more barely centered than the general cost-related things that are incorporated into the main set. Work related issues incorporate troubles that emerge in light of the fact that new advancements by and large need higher aptitude levels of workers. Even with these requirements, a firm may experience obstacles to implementation of advanced manufacturing systems. if there is a deficiency of aptitudes accessible on the market, or on the off chance that it confronts training problems in overcoming inadequacies, or if its work contracts act to compel its capacity to substitute labor crosswise over errands.

Authoritative issues are those related with troubles in executing the sorts of inside changes in a firm that are required for the use of advanced manufacturing systems. The first of these is the trouble in acquainting imperative changes to the company. For instance, the presentation of computer-supported design may require new structure that connection building improvement with the manufacturing control office so that the benefits of concurrent engineering can be completely used. Other authoritative issues come from a poor state of mind of senior administration towards new advancements, or labor resistance. Establishment / government related issues are those related with charge administrations (both the R&D impose credit and capital cost recompenses) and with government directions and norms. Data and info related issues emerge if markets for learning are blemished. They encompass poor logical and specialized information, innovative administrations and specialized technical from contractors.

THE POTENTIAL OF INFORMATION TECHNOLOGY IN MANUFACTURING

A gigantic measure of data is created and utilized during the planning, design, production, and utilization of a product to fulfill client needs and to meet environmental prerequisites. In this way it is reasonable to assume that the utilization of information technology can empower generous upgrades in the operation, association, and viability of computer based intelligent manufacturing processes and exercises, generally by encouraging their joining with each other resulting in an integrated manufacturing setup. Equipment and stations inside processing plants, whole manufacturing factories, and systems of providers, partners, and clients situated all through the world can be all the more viably associated and coordinated using information technology.

Information technology can give the instruments to enable ventures to accomplish objectives generally viewed as basic to the eventual fate of manufacturing, including:

- Rapid moves in production with one product then onto the next;
- Faster execution of new ideas in products;
- Faster conveyance of products to clients;
- More close and itemized cooperation with clients;
- Fuller usage of cash and HR;
- Streamlining of operations to concentrate on basic business needs; and
- Elimination of pointless, excess, or wasteful exercises.

The improvement and execution of new information technology to meet these objectives is formed by hierarchical, administrative, and human asset needs that have kept producers from exploring completely even the innovation that exists today. (Beatty, n.d.)Affectability to these needs will be fundamental to the effective improvement and execution of the information technology associated with dreams of manufacturing. Information technology can be utilized to meet a scope of necessities of manufacturing heads. These requirements propose an exploration motivation with both mechanical and non-innovative measurements; the essential focuses of this examination plan include:

- Operational control of plants and their providers,
- Tools for product and process plan,
- Modeling and reenactment of the whole range of processing plant operations (virtual manufacturing), and Enterprise

incorporation and utilization of different capacities given by the developing National Information Infrastructure to help modern manufacturing.

SMART MANUFACTURING

Smart manufacturing weds information, innovation, technology and human inventiveness to realize a quick insurgency in the advancement and use of manufacturing knowledge to each aspect of business. It will on a very basic level change how products are sorted, made, dispatched and sold. It will enhance laborer well-being and secure the environment by making zero emissions, zero-accident manufacturing conceivable. It will help keep employments by keeping makers focused in the worldwide marketplace regardless of the generously higher cost of working together.

Smart manufacturing advancements drive these changes in three fast dynamic phases:

Phase 1: Manufacturing Plant and firm's business wide Incorporation

The mix of all manufacturing data throughout individual manufacturing units and crosswise over many firms encourages significant, quick upgrades in costs, wellbeing and environmental effects. In its first step, smart manufacturing interconnects and better corresponds various phases of manufacturing to progress the efficiency of the system. An average manufacturing plant utilizes information technology, automated controls, production administration programming, sensors and the like to deal with each specific stage or operation of a manufacturing process. In any case, each stage of manufacturing is a part of efficiency. Smart manufacturing incorporates these parts of efficiency, empowering information sharing all through the plant. The joining between machine-accumulated information and human knowledge will propel expansive enhancement and venture wide administration destinations, incorporating generous increases in financial execution, specialist security and environmental manageability. The development of this "manufacturing knowledge" will introduce the second phase of smart manufacturing.

Phase 2: From extensive Optimization to Manufacturing Intelligence

The information, combined with cutting edge PC recreation and demonstrating, will make strong "manufacturing insight" that will empower variable-speed, flexible manufacturing, ideal production rates, and faster product customization. Associating manufacturing plant modeling demonstrations and information technologies with elite registering and computing platforms (distributed computing) makes it conceivable to assemble significantly larger amounts of manufacturing knowledge and interface it all through the industrial facility. Finish production lines and whole plants run at variable paces with ongoing flexibility — which was not feasible before — with a specific end goal to monitor energy usage and increase yields. Organizations have the capacity to create advanced models and recreations of manufacturing processes to enhance present and future operations. For example, producers have the capacity to create models for the mass-manufacturing of products and gadgets that utilize nanotechnology. Nanotechnology, as the name infers, includes the advancement of ultraminiaturized, very intricate gadgets, systems and materials. It is broadly altering technologies, innovations and industry with littler, more smaller, lighter weight materials and capable precise gadgets for about each industry. Screens as little as the leader of a pin might be embedded in a patient's blood vessel stent to gather continuous information on blood flow. Lighter weight vehicles, produced using nanotechnology-empowered materials, are more fuel-efficient and more secure. Nanotech-empowered sensors, custom fitted to distinguish specific poisons and biohazards, are helping in protection of our military, police and the general population. It gives the idea that nanotechnology is having even more significant effects on each aspect of our lives than did the creation of microchips. Its usage requires the exact modeling and manufacturing that smart manufacturing will empower.

The second phase of smart manufacturing too will associate manufacturing plant specific data to information all through the network — from crude material accessibility and client demand through the conveyance of finished products. It will encourage the utilization of keen lattices to plan energy-concentrated exercises amid low-demand periods and moderate production amid crest energy demands. It will empower more prominent product customization, new product reenactments and more efficient processes. It will bolster the production of more secure products and exactly defend, faster tracking of the products. (Beatty, n.d.)

Phase 3: Manufacturing Knowledge

As that manufacturing insight develops, it motivates developments in processes and products that involve smart manufacturing's guarantee a noteworthy market disturbance. Manufacturing Knowledge is greatly changing the manufacturing Markets. smart manufacturing's third phase is changing industry comparably to how the vital utilization of information technology changed the plan of action — and buyer shopping conduct. Amazon started 15 years back as an online book retailer. By catching broad information about customers' book-purchasing propensities, Amazon created broad information about its clients' ways of life that empowered a "problematic" move in its plan of action. In 10 years, Amazon expanded its product offerings to traverse numerous new classes. It is presently the biggest U.S. - based online retailer. Books and other media speak to just 52 percent of its $24 billion dollar net deals. As any individual who has requested from Amazon knows, its Web webpage will propose products you may like and reveal to you what different shoppers purchased when they considered a product you are seeing. To put it plainly, Amazon makes smart utilization of its industry information to support its upper hand. Smart manufacturing conveys a comparable, remarkable move in the aggressive landscape of manufacturing industry. It is switching the flow of the 100-year-old modern supply chains that constrained clients to acknowledge whatever was mass created. Flexible processing plants and IT-advanced supply chains are changing manufacturing processes to permit makers to modify products to person needs, such as drugs with specific measurements and plans. Clients will «tell» a processing plant what product to make, what components to assemble into a PC or how to tailor a couple of pants for an impeccable fit. This most sensational and important third phase of actualizing smart manufacturing originates from development impelled by this developing collection of manufacturing information. These are not incremental or progressive changes, they are competition changer, radical developments in products and processes. Changes at this phase push down costs, open new markets and offer a more extensive exhibit of decisions to a more extensive scope of individuals.

ADVANCED PLANNING AND SCHEDULING

It is advanced planning, arrangement and scheduling of materials and booking (otherwise called APS or advanced manufacturing) it alludes to a manufacturing administration control by which crude materials and production limit are ideally assigned to meet demand. APS is particularly appropriate to environments where less difficult arranging strategies can't sufficiently address complex exchange offs between contending needs. Production planning is inherently extremely troublesome due to the (roughly) factorial reliance of the extent of the arrangement space on the quantity of things/products to be fabricated. (Bessant, n.d.)

Difficulties of Production Planning

Customary production scheduling and booking systems (such as manufacturing asset arranging) utilize a step by step methodology to designate material and production limit. This approach is basic yet unwieldy, and does not promptly adjust to changes in demand, asset limit or material accessibility. Materials and limit are arranged independently, and numerous systems don't consider material or limit imperatives, prompting infeasible plans. Therefore, endeavors to change to the new system have not generally been effective, which require the blend of administration theory with manufacturing.

Not at all like past systems, APS at the same time designs and timetables production based on accessible materials, have work and plant limited.

APS has normally been connected where at least one of the accompanying conditions are available:

- Make to arrange (as unmistakable from make to stock) manufacturing
- Capital-escalated production processes, where plant limit is obliged
- Products "contending" for plant limit: where a wide range of products are delivered in every part
- Products that require many manufacturing tasks
- Production requires common plan changes which can't be anticipated before the occasion

Advanced planning & scheduling programming empowers manufacturing booking and propelled planning improvement inside these environments (Bessant, n.d.)

The Manufacturing Planning and Control Processes

Practically many organizations design and control their material and production streams in a various leveled structures with various timelines and itemized data. One hierarchal arranging setup generally utilized is named as Manufacturing planning and control (MPC) system.

The MPC processes chain of command was started from the Association of operation Management where deals and operations arranging and has the longest arranging line and the littlest level of detail. It adjusts the deals/promoting plans with present production assets and results in an conceded to organization blueprint that decides the manufacturing part in meeting organization methodology. Next takes after master production scheduling-MPS. It is the non-accumulated variant of the deals and operations scheduling. Asset planning decides the limit important to deliver the required products now and in the future. There are numerous similarities between deals and operations scheduling and master production scheduling and in many organizations these both arranging levels are incorporated into one scheduling and arranging process. Next in the chain of command comes the order scheduling and planning, which could be viewed as the capacity inside an organization that executes plans set up at the vital and strategic level at the organization. The last procedure in the master production scheduling system is the production steps or activity control with the most limited arranging and planning limit and the biggest level of detail. (Bessant, n.d.)It concerns the execution of nitty gritty material designs and depicts the arranging and release of individual requests/orders, and when fundamental point by point planning and control of individual occupations at work focuses. Arranging in understanding with the master production scheduling system is once in a while called manufacturing asset arranging or manufacturing resource planning. As the fundamental concentration in the proposition is sales and operations planning and production activity control.

Sales and Operation Planning

Considering the way that S&OP have existed as a rule for no less than 25 years, moderately little has been investigated up to this point. Some say that the acknowledgment is on the ascent and that S&OP will increase in significance as its unpredictability and the rate of progress increases over the business. Others recommend that organizations are moving far from S&OP as they have noted that it does not satisfy the requirements. Up to this point, research into on S&OP has concentrated on its definition, processes, exercises, usage methodology, and case ponders tending to the banquet after its execution. As indicated by the word reference of APICS the meaning of S&OP is "setting the general level of manufacturing yield and different exercises to best fulfill the current arranged levels of offers while meeting general business objectivities of gainfulness, productivity etc., as communicated in the whole business plan for success". Similarly as the definition suggests, S&OP sets the casing for the choices at the lower levels, based on the marketable strategies, business objectives and future dreams. It is the primary connection between the top administration and the subordinated plans. Choices taken at this level are of the best effect on benefit and intensity. S&OP tries to adjust different capacities in the organization and is some of the time seen as a device that enables the figuring out how to group towards aggregate basic leadership. Commonly the S&OP is made on an "accumulated" or "family" level and covers an adequate traverse of time to ensure that vital assets are accessible. Generally, S&OP can be partitioned into a business design and a production design, which decides the limit necessities, stock level and/or accumulation level. The limit planning at the S&OP level concerns the assessment of limit necessity, choices about future limit modifications, and the issue with having adequate limit is of a money related nature. There are two basic procedures used to build up a production design at the S&OP level; chase and level. A chase technique implies that production is changed to coordinate demand, whereas a level technique delivers a sum equivalent to the normal demand. By and large it is definitely not conceivable (nor attractive) to keep up an unadulterated S&OP arranging procedure. Rather, a center route with the point of finding an effective exchange off between the unadulterated methodologies is connected. At the point when overseen in a proper way the S&OP procedure should produce many advantages to the organization in the type of enhanced client benefit, decreased stock levels, shorted client lead times, balanced out production rates, better work with providers, moved forward collaboration between deals, operations, back and product improvement, as well as potential outcomes to respond to new business openings (Bessant, n.d.)

S&OP regularly takes after a five-stage procedure;

- Predicting the demand of products in future
- Setting up a preparatory delivery schedule and defining up objectives for stock or client request logs.
- Preparing a preparatory manufacturing schedule.
- Changing the delivery schedule design and manufacturing schedule.
- Resolve the delivery schedule and manufacturing schedule.

How regularly the S&OP procedure is directed fluctuates from case to case. It likewise relies upon the kind of business, present lead times of delivery, how quickly the market values alter and the recurrence of product restoration or renewal. Another factor to think about when directing the S&OP procedure is the number of times when it is important to cross match different manufacturing operations with their financial plans and to make new spending forecasts. (H. R. Parsaei, n.d.) As a rule, sales and operations scheduling processes are completed month to month yet can likewise be done all the more as often as possible. Programming systems might be required in the S&OP procedure, yet not at very start of schedule in the implementing procedure. The reason is that it is more essential to have a strongly planned S&OP business control rather than having a rich programming system. There can be an S&OP combination structure to sort the development level of various S&OP processes. The system utilizes a one to five positioning crosswise over five areas; gatherings and coordinated effort, association, measurement, information technology and S&OP design and planning incorporation. A Stage 1 organization does not have any S&OP plan or control, whereas Stage 5 identifies an organization with a proactive S&OP procedure portrayed by occasion - driven gatherings, constant access to outer information, consistent combination of plans, and benefit - centered processes. A Stage 5 organization is likewise an organization where S&OP is comprehended to be an instrument for enhancement. Operations enhancement programming, such as limited planning and scheduling and continuous coordinated arrangements with S&OP workbench for sharing data about deals and manufacturing operations planning among colleagues, are in all probability required to accomplish a Stage 5 S&OP procedure. The awesome capability of Advanced planning and scheduling systems provide broad support to the S&OP procedure since its abilities of incessant rescheduling prompts a few changes in the S&OP choices, in this way lessening the dependability of the plans at the operational level and all through the inventory network. It is essential to utilize decision help in the S&OP procedure since it is hard to achieve accord among the distinctive offices and amid gatherings.

Production Activity Control

Production activity control is an approach for implementing the material handling plans and has three fundamental purposes

- To release orders at the rate that limit conditions will enable them to be executed with reasonable throughput times
- To guarantee that starting materials are accessible when each request or order is wanted to begin
- To guarantee that requests released for manufacturing in the processing shop are finished in a reasonable time frame concerning delivery exactness and throughput times.
- The Production activity control procedure comprises of various exercises exhibited in Figure. The relative significance of these exercises to a substantial degree relies upon the manufacturing processes of the organization. The essential contrasts are between a product line and a project setup control. In a product line the control of request release and need control happens at the same time. In manufacturing place, then again, all exercises in Production activity control are generally equivalent in significance and degree. As a rule the following applies: First the upper planning and arranging levels start a manufacturing request; the request ordinarily contains data on begin time, due date and amounts. (H. R. Parsaei, n.d.)

Prior to the request is released to the workshop there must be capacity and capability available to complete it (arrange release control), material vital for its making must be accessible (material accessible check) and data required for executing the manufacturing request must be conveyed to the workshop (producing a shop bundle). At the point when the request is released to the workshop, its operations must be executed in a proper succession. This action is called sequential manufacturing or need control. The execution of the operations in the workshop is frequently alluded to as dispatching. At long last, the way that manufacturing orders and operations have begun and complete, circumstances does not imply that they fundamentally will begin or finish at the times mentioned. Subsequently, data relating to the advance of requests in the workshop ought to be accounted for to the higher planning and scheduling levels with the goal that they can know about what is going on as well as mediate in request to rectify any possible issues. There are three distinct levels comprising work reporting: to report the whole request, to report the operations and to report materials pulled back for the request or conveyed when the request has been finished. (Bessant, n.d.) (H. R. Parsaei, n.d.)

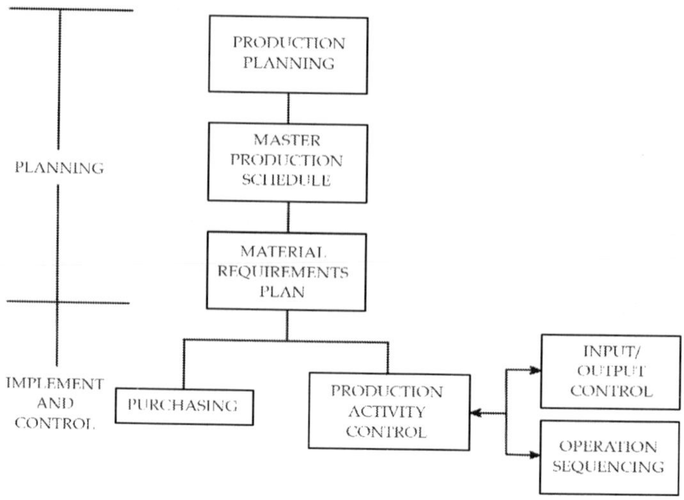

Figure 11 production activity control and priority planning (Anon., n.d.)

Most of the manufacturing organizations today utilize MRP rationale in ERP systems to schedule material necessities and release manufacturing and purchasing orders. By and by, there are a few techniques to guarantee that the released requests to the manufacturing shop are executed in a suitable time frame. A few organizations pass the rundown of released requests to the managers or foremen, who settle on small scheduling choices, bolstered by experience of the person or, then again basic priority rules. Other organizations make utilization of pretty much modern planning calculation in PC systems. As people are not very much prepared to control or advance substantial and complex systems, specialists in manufacturing scheduling and control are frequently persuaded that much can be enhanced with manual planning. (H. R. Parsaei, n.d.)The scholarly field in manufacturing planning, scheduling and control has for a considerable length of time defined different planning methods while trying to render feasible production booking. Production planning issues are generally classified as single machine issues, stream shop issues, and project shop issues. Distinctive target capacities and extra criteria like needs, sequence based setup times or parallel assets prompt an enormous number of planning issue classes. For each class of planning issues, straightforward need rules, and moreover refined scheduling calculations have been produced.

CHAPTER 2

Forming Processes

INTRODUCTION

Metal forming is an imperative manufacturing operation. It has commercial significance among different production operations because of its focal points such as cost viability, upgraded mechanical properties, flexible operations, higher productivity, impressive material sparing.

The items and articles that we use in our day by day life are man-made, built parts, which are acquired from some crude material through some manufacturing procedure. All these products are made of various little segments assembled into a product. The pen that we use for composing, for instance is made of a few little parts, assembled together. A car should be an assembly of more than 15000 parts, created through different manufacturing operations.

Manufacturing of completed parts and segments from crude materials is a standout amongst the most vital strides in production. Production encompasses a wide range of manufacturing processes. Manufacturing alludes to the transformation of crude materials into completed products utilizing reasonable strategies. There are many techniques for manufacturing such as metal casting, metal forming, metal machining and metal joining.

A portion of the present day techniques for manufacturing incorporate small scale machining, nano- production, ultra accuracy manufacturing etc. Keeping in mind the end goal to satisfy the prerequisites of the continually increasing demands of different sorts of ventures, the manufacturing engineer has to pick the correct sort of material and the right kind of equipment for manufacture with the goal that the cost of production and the energy utilization are least.

The determination of reasonable manufacturing procedure should be selected after evaluating it for environmental effects such as air contamination, waste disposal etc. Present day ideas such as lean manufacturing, versatile control, agile manufacturing, group and cell technology etc have extensive impact on cost decrease and quality enhancements of products.

Computers and robots assume essential part in present day manufacturing strategies, today. Modeling and Demonstration of the products and procedures before mass production helps the manufacturing engineer select the best working parameters and thus accomplish the completed product to the most extreme level of value and cost-adequacy.

SUPER PLASTIC FORMING

Super plastic forming is a metalworking practice for making sheet metal. It is based on the theory of super plasticity, which states that " some materials have ability to bear large and uniform deformation before necking and fracturing". Super plastic forming (SPF) of sheet metal is used for making products with complex shapes and integrated parts which are often more light weight and stronger than the assemblies made by other processes. (Giuliano, 2011)

Some best examples of materials with super plasticity are:

- Aluminum alloys
- Zinc-aluminum alloys
- Aluminum-lithium alloys
- Titanium alloys
- Bismuth-tin alloys

- Stainless steel

Super plastic forming is a proximate shaping process that usually employs high temperature and gaseous pressure to shape materials by using a single step mold. Super plastic materials are able to bear very high elongations of about 200% or more. (Eric M. Taleff. St, 2004)A material shows super plastic behavior if it follows the criteria published in Metal Forming Mechanics and Metallurgy Second Edition (6). The criteria are stated below:

- Grain size is very fine of a few microns or less, with identical and equiaxed grains structure.
- The temperature is high - usually about half of the melting temperature of that material.
- Strain rates are low on the order of 10-2/sec or even lower than this.

Super plastic forming (SPF) can produce those parts that are very difficult to produce by using conventional available techniques. In the SPF process, the material is enclosed in a closed die and is heated to the SPF temperature. After that Inert gas pressure is applied, at a specific and controlled rate to force the material to adopt the shape of the die pattern. Super plastic alloys have the characteristics of stretching at higher temperatures by many fold of their original length without any rupture. (Eric M. Taleff. St, 2004)

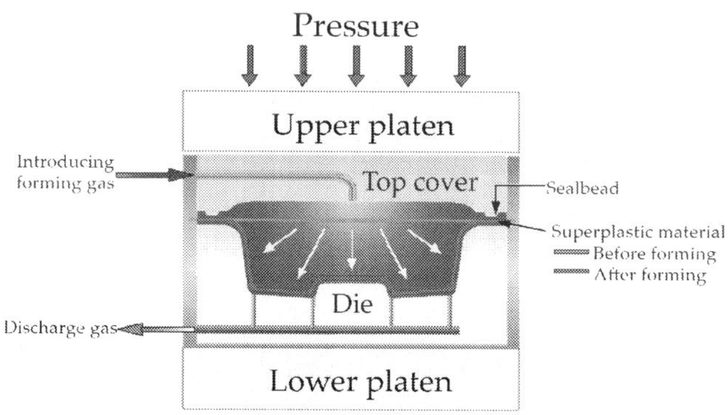

***Figure 2*1**Superplastic forming process description along with equipment used (http://mechanicaluniversity.blogspot.com)

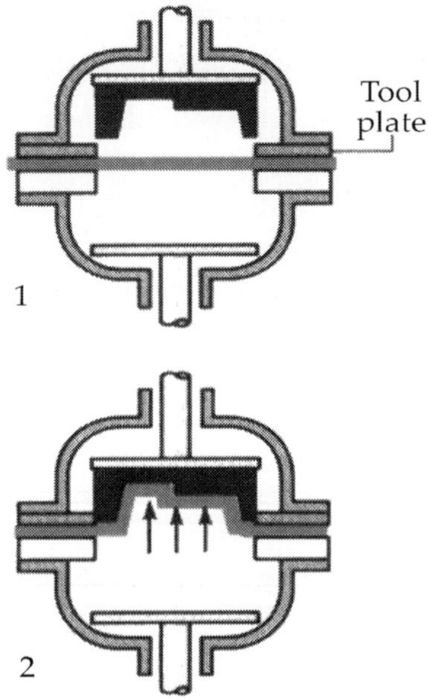

Figure 22 Female super plastic forming using dies and force (Anon., n.d.)

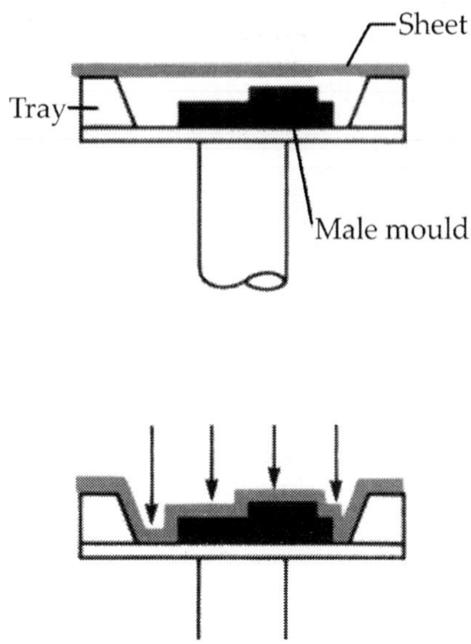

Figure 23 Female drop super plastic forming using dies and force (Princeton)

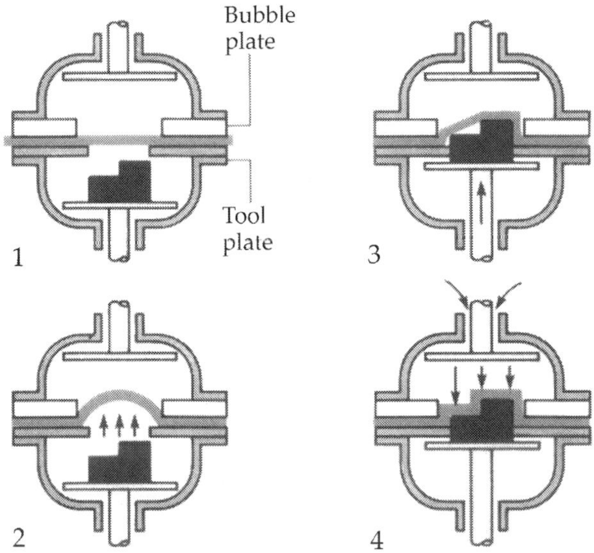

Figure 24Plug-assisted snap back super plastic male forming process (Princeton)

Graphite-coated blank material is placed in the heated press. The Blank is formed into the shape of bubble and male mould is pressed into the blank bubble. Air pressure is used for forcing the metal in precise contact with the mould surface.

There are many gains and shortcomings to using super plastic forming above other forming processes.

Examples of the applications are

- Fuel Tanks
- Fuselage of aircrafts
- Motorcar parts

Advantages

- The end product has outstanding precision and a well surface texture.
- Larger products can be produced by altering the design for eliminating assemblies and reducing weight, that is a crucial consideration in aerospace industry.

- The work piece required need not to be very strong which keeps tooling costs low.
- The capability to produce multifarious parts by large deformations.
- Cost of die and tooling is low die which facilitates cost-effective product customization.
- Spring back is reduced which make the process safe.

Disadvantages

- Low production rate.
- Work piece must not become super plastic at Lower temperatures. It should behave as strong material at lower temperatures so that it can be properly handled.
- It is typically employed for making less number of products.
- Pre-forming steps are expensive and time consuming. Materials are refined and heat treated before the super plastic forming so that they can exhibit the super plastic behavior.
- Products show poor creep behavior.

HIGH ENERGY RATE FORMING (HERF)

Today's manufacturing industry demands superior quality products with excellent productivity and greater economy. In order to meet these challenges a smooth and trouble free high energy rate forming technique is of special importance. (Company, 2004)

High energy rate forming (HERF) is a forming process which uses huge amount of energy rapidly for shaping the materials. HERF processes offer many advantages over conventional forming process and this is the principal reason behind their increasing popularity. These processes offer greater control, precision, manufacturing speed and ease of complex shape production which produce environment friendly and economic products. (Benedict, 1987)

Decreased weight is one of the main targets in the automotive and aerospace industry. HERF processes prove to be very promising for the forming of light weight alloys such as Aluminum alloys which have lower density and high strength to weight ratio, high ductility and excellent corrosive resistance. (Davis, 1995)

Traditional forming processes do not take inertia of the materials under consideration because speed of forming is mostly less than 5 ms-1, on the other hand, usual high energy rate forming processes are performed at work-piece speeds of the order of 100 m/s. Many methods of HERF are available these methods differ on the basis of source of energy used for forming. Frequently used methods for HERF include

- Explosive forming
- Electro hydraulic forming (EHF)
- Electromagnetic forming (EMF).

Amongst these methods, electromagnetic forming is a high-speed forming method, which uses pulses of magnetic field to form the work piece geometry. Electromagnetic forming uses conductive metals for example copper and aluminum alloys. The products formed by electromagnetic forming offer higher hardness, less corrosion rate and good precision. (Company, 2004)

In explosive forming process the high energy produced by explosion of an explosive substance is used for forming the metal sheets. No punch or hammer is needed only a hollow die is employed. The sheet metal is fastened on the upper side of die and the hollow space below the sheet is evacuated by using vacuum pump. The whole arrangement is positioned in a water-filled tank. The explosive material which is placed at a certain distance from the die is then fired. This explosion generates shock waves. (Davis, 1995) The maximum pressure produced due to the shock wave is calculated by the following equation:

$$p = k\left(\sqrt[3]{w} / R\right)^{a}$$

k is a constant of proportionality, a is material and process constant.

R represents the stand-off distance.

Maximum pressure developed by the explosive also depends on the Compressibility and impedance of the medium. If the compressibility of the medium is higher, the maximum pressure or peak pressure developed would be low. If medium has higher density the maximum pressure or peak pressure of the shock-wave would be higher. (Benedict, 1987) Explosion speeds as much as 6500 ms-1 is frequently used. The material flow is also very fast in explosive forming of the order of about 200 ms-

1. Greater material flow means higher strain rates. Only those Materials which remain ductile at higher strain rates are used for explosive forming. The stand off distance also plays important role in determining the peak pressure produced by explosive forming. Steel plates of up to 25 mm thickness can be formed by explosive forming. Tubes are also bulged using explosive forming.

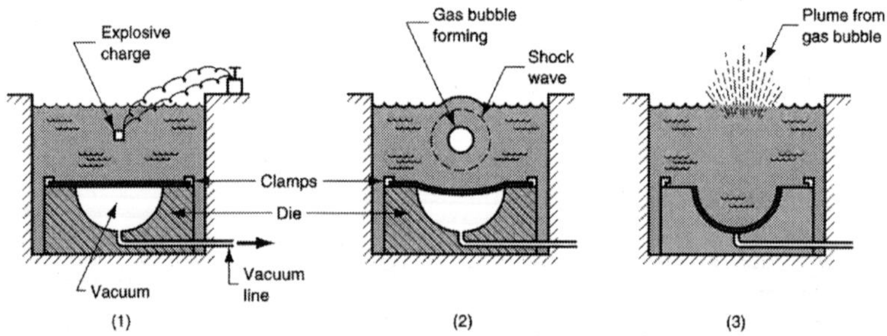

Figure 25 Description of explosive forming process - a kind of high energy rate forming process (Anon., n.d.)

The efficiency of High energy rate forming processes is highly dependent on strain rates. Strain rates affect product properties in many ways during forming:

- As strain rate increases flow stress to continue the deformation increases.
- At higher strain rates the temperature of work piece is increased because of adiabatic heating.
- Higher strain rates need improved lubrication.
- Many materials which are difficult to form by conventional forming processes can be formed by high strain rate processes e.g. Titanium and Tungsten alloys.

Principle / Important Features of HERF Processes:

- Rate of energy supply to work piece is much greater than the conventional process.
- Energy is supplied for a very short time interval.
- Material speed during high energy rate forming is higher than conventional forming processes, so these processes are also

termed as high velocity forming processes.

- Many materials form quickly under rapid action of stresses.
- Larger parts can be conveniently manufactured by using these processes.
- For many metals ductility displays regular pattern, it increases with strain rate upto a certain limit and then decreases. To form a material by high energy rate forming processes it is strained up to the safe processing limit.
- As temperature increases dependence of strength on strain rate increases.
- At low plastic strains flow and yield stresses are more dependent on strain rate than the dependence of tensile strength on strain rate.
- Due to higher strain rates, yield point appears in low carbon steel which do not show a yield point under normal strain rates.

Advantages of HERF Processes

- Rapid production means higher production rates.
- Relatively lower die cost.
- Tolerances are easily retained.
- Process is versatile and it can form most metals even difficult to form metals can be formed by this process.
- After the forming process there is very minimal spring back effect on the work piece.
- The cost of Production is low because large presses and mills are eliminated in this process which makes the process economically justifiable.
- Complex profiles are easily formed by High energy rate forming processes as compared to conventional forming processes.
- There are no pre forming and intermediate forming steps in the forming processes and desired product is produced in a single step.
- The process is effectively suitable for a range of production volumes e.g. small number of products, batches and large quantity production.

Limitations

- At all stages of the HERF process, highly skilled personnel are needed.
- Work piece has to bear many transient stresses of very high magnitude.
- HERF processes are unsuitable to many brittle materials..
- Energy sources need careful handling.
- As the process uses more sophisticated technologies and explosives, Governmental permissions should be taken and safe work procedures must be followed.
- Dies should be more tough and resilient to withstand rapid energy rates and bursts to avoid defects.
- The amount and way of delivering energy to the work piece is critical and excessive energy delivery can damage the die or work piece.
- The behavior of the material which is to be processed should be properly established before processing.

Applications

- Ship industry for making large parts and plates.
- Bending tubes and pipes of high thickness. Crimping the metallic strips and tubes.
- Forming dishes for radars.
- Space applications e.g. elliptical domes.
- Metallic cladding of dissimilar metals

ELECTROMAGNETIC FORMING

It is a kind of high speed cold forming process which is used for electrically conducting materials/metals most frequently aluminum or copper. This process is also termed as magnetic pulse forming, and is mostly employed for swaging processes, for example, clipping fittings on the tube ends and crimping the cable ends. Many other uses of the electromagnetic Forming are embossing, drawing , blanking and blanking. (Benedict, 1987)

The principle of electromagnetic forming is that the material is placed inside a coil. There is a bank of parallel capacitors which are supplied a high voltage for charging. Using more capacitors or by using the charging voltage, amount of stored electrical energy can be increased.

After completing the charging, a switch is used to deliver this electrical energy to the coil which establishes a strong magnetic field. This magnetic field produces straw currents in the work piece. These straw currents also produce a magnetic field. (Hyeon Il Park, 2014)

The forces resulting from the two magnetic fields act in opposite direction to each other and a repelling force is established between the coil and the work piece. This force cause work piece deformation. We can use permanent coils or expandable coils. As the coils also bear repelling forces, it must have enough strength and strong insulation to withstand these forces. The expandable coils are much cheap, and are used when high energy levels are required.

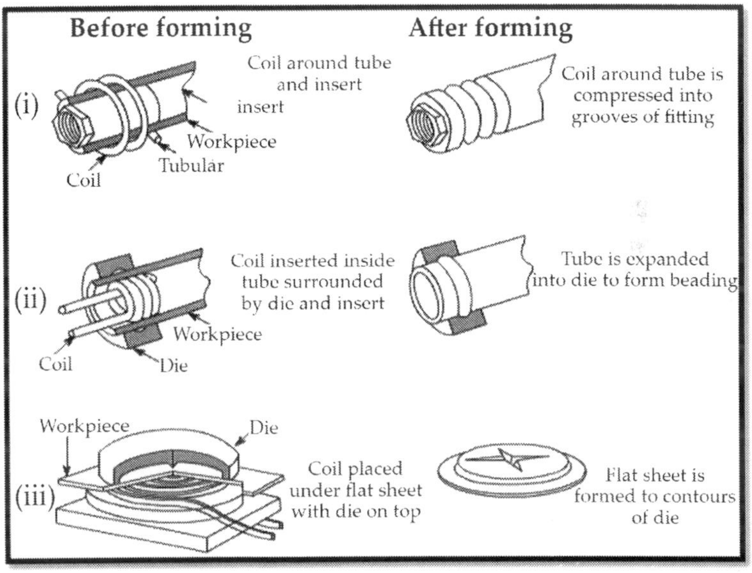

Figure 26Various applications of electromagnetic forming process (i) Compression (ii) expansion (iii) sheet metal forming (Anon., n.d.)

Electro Magnetic forming can use different types of below mentioned coils, subject to the job and necessities. (Hyeon Il Park, 2014) Ring compression coils. The coil is like expansion coil and for performing the forming operation it is placed around the tube which is required to be compressed. As shown in above Figure (i)

Tube expansion coils. The coil is set in the tube which is required to be expanded. As shown in above Figure (ii)

Flat coil. It is a spirally wound metal strip in a certain plane. These coils are employed for sheet metal processing.As shown in above Figure . (iii).

The electromagnetic forming technology has distinctive benefits in the production of light weight metals due to enhanced

- Enhanced formability
- Superior mechanical properties
- Enhanced strain distribution
- Reduced wrinkls,
- Active spring back control
- Minimum distortion of local structures and die.

The most common uses of electromagnetic tube compression process are

Shape joints. To join metallic tubes with internal metal mandrels which are used for axial and torsional loading applications.

Friction joints. To join metallic tubes with non-metallic internal mandrel e.g. wire rope.

Solid state welding. To join tubes and internal mandrels of different metals e.g. chassis components, tow poles and aircraft torque tubes.

The electromagnetic forming process has many potential advantages above conservative forming processes. Main advantages are listed below:

- Excellent ability to produce many shapes.
- Improved wrinkle elimination.
- This process can be used in combination with other manufacturing processes easily even for joining dissimilar parts e.g. plastic, glass and composite materials.

- High precision and dimensional tolerances are achieved due to less as spring back.
- Single sided dies decrease the cost of tooling.
- Almost no lubricants are required, so this process can be done in clean enviorment.
- Excellent reproducibility. Control over the current supplied is key to good forming process.
- Improved surface finish of the work piece because there is no contact between die and work.
- High production rate.

The practical coil is suitably designed for dealing with varying job requirements of every forming process.

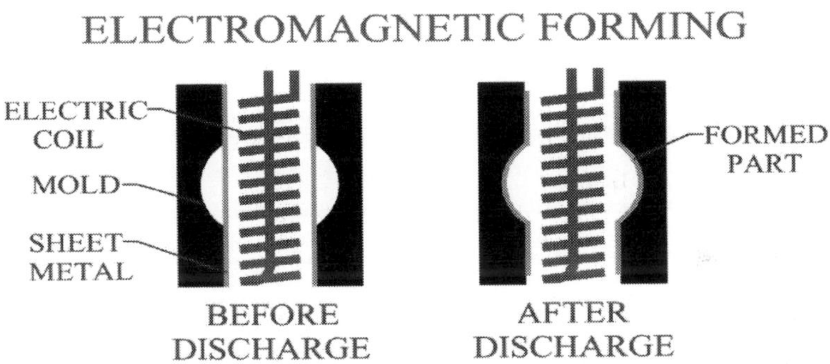

Figure 27 Illustration of Electro Magnetic Forming (Anon., n.d.)

Process Details/ Steps:

- The capacitor bank stores the electrical energy.
- The tubular work piece is attached on a mandrel which has the cavity which is used to produce shape on the tube.
- A primary coil is positioned about the tube and mandrel assembly.
- When the switch is closed, the energy is discharged through the coil

- The coil creates a changing magnetic field about it.
- The secondary current induced in the tube generates its magnetic field in the opposite direction to the parent magnetic field.
- These magnetic fields act in opposite direction to each other and firmly gripped coil repels the work piece in the die.
- The work piece tube sets in the die and takes its shape.

Process Parameters

- Size of work piece
- Electrical conductivity of the work piece metal.
- Capacitance of capacitors bank
- The intensity of current, which determines amount of magnetic field and applied force.
- Insulation of coils
- Rigidity of coils
- Limitations:
- Only useable for electrically conducting materials.
- Un- suitable for large products.
- Firm clamping of the coil is sensitive.
- Coil life is shot because of huge forces which are acting on it.

Applications

- Crimping
- Bending
- Bulging

FLOW FORMING

Flow forming is a modernized, enhanced propelled variant of metal spinning, which is one of the most established techniques for chipless forming. The metal spinning technique utilized a rotated pointer to physically push a metal sheet mounted toward one side of a spinning mandrel. This strategy was used to manufacture axisymmetric, thin walled, light weight household items, for example, pans and cooking pots. (Benedict, 1987)

Flow forming is a procedure whereby a metal blank, a plate or a hollow tube are mounted on a mandrel which pivots the material to make flow pivotally by at least one roller along the pivoting mandrel.

The significant contrast amongst spinning and flow forming is, in spinning, the thickness decrease is exceptionally minor and in flow forming the variety in thickness can be retained at some places along axial directions. Flow forming implies forming a sheet metal, tube or draw piece in one or more passes of the forming roll or rolls. The amount of wall thickness reduction relies on the properties of the material and the number of passes. (Taylan Altan, 2012)

Flow Forming is an incremental metal forming method in which a plate or tube of metal is framed over a mandrel by at least one rollers utilizing enormous pressure. The roller stresses the workpiece, compelling it against the mandrel, both axially increasing and radially thinning it.

Since the pressure applied by the roller is very restricted and the material is incrementally shaped, frequently there is a net funds in vitality in forming over drawing processes. Flow forming subjects the workpiece to a lot of friction and distortion. These two variables may warm the work piece to a few hundred degrees if appropriate cooling liquid is not used. Flow forming is frequently used to manufacture automobile wheels. Amid flow forming, the workpiece is cool worked, changing its mechanical properties, so its strength ends up noticeably like that of forged metal. Flow forming, otherwise called tube spinning, is one of the systems firmly associated to shear forming. (R.L. Timings, 2009)

The distinction is as per the direction of material flow and direction of movement of tool (roller). In the event that both are in same direction, it is forward flow forming and on the off chance that they are in inverse direction, it is backward flow forming. Forward flow forming is reasonable for long, high precision thin walled segments. Backward flow forming is reasonable for blanks without base or internal flange. In forward spinning the roller moves far from the settled end of the work piece, and the work metal flows in an indistinguishable direction from the roller, typically toward the headstock. The primary favorable position in forward spinning when contrasted with backward spinning is that forward spinning will overcome the issue of distortion like bell-mouthing at the free end of the blank and loss of straightness. (Taylan Altan, 2012) In forward spinning more control of length is possible on the grounds that as metal is framed under the rollers it is not required to move again and any variety caused by the variable wall thickness of the preform is constantly pushed ahead of rollers, in the long run getting to be noticeably trim metal past the completed length. The impediment of forward flow forming is that the Production is slower in forward spinning on the grounds that the roller must transverse the completed length of the work piece.

In backward flow forming the mandrel is unsupported. In backward spinning the work piece is held against an apparatus on the head stock, the roller progresses towards the settled end of the work piece, work flows the other way. The benefit of backward flow forming over forward flow forming:

The preform is more straightforward for backward spinning because it slides over the mandrel and does not require an internal flange for clamping.

The roller transverse just 50% of the length of the finished tube in making a decrease of 50% wall thickness and just 25% of the final, for a 75% decrease. We can handle 3 m length tube by utilizing mandrel.

In both the flow forming processes, there is no distinction in stress and strain rate.

The major disadvantage of backward tube spinning is that backward flow forming is normally inclined to non uniform measurement across the length of the work piece. In this Process as appeared in Fig. a, the metal is displaced axially along a mandrel, while the internal diameter remains constant. It is usually utilized to deliver cylindrical parts. Most present day flow forming machines utilize a few rollers and their design is more difficult compared to that of spinning and shear forming machines. The starting blank can be as a sleeve or cup. Blanks can be delivered by deep drawing or forging in addition to machining to enhance the dimensional accuracy. (Benedict, 1987) Advantages, for example, an increase in hardness because of an ability to cold work and better surface finish couples with basic tool design and tooling cost make flow forming a particularly attractive procedure for the creation of hydraulic cylinders, and cylindrical hollow parts with various ventured segments. (R.L. Timings, 2009)

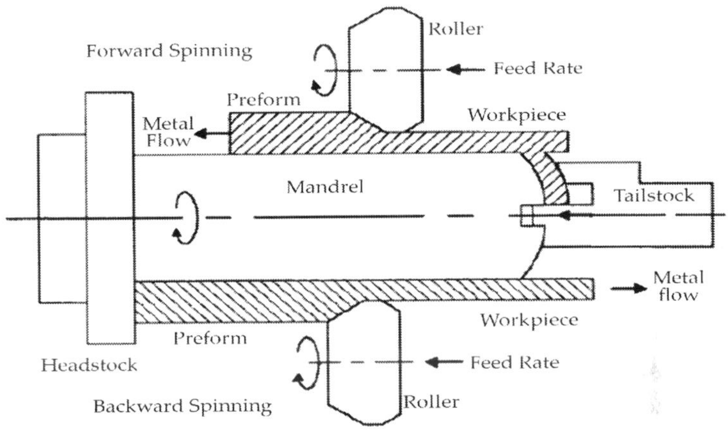

Figure 28 Illustration of Forward & Backward Flow Forming processes (Anon., n.d.)

In flow forming, as demonstrated schematically in Fig. a, the blank is fitted into the rotating mandrel and the rollers approach the blank in the axial direction and deform the metal under the contact point. Along these lines, the wall thickness is lessened as material is encouraged to flow mainly in the axial direction, increasing the length of the workpiece the final segment length can be calculated as,

$$L1 = \quad L0\frac{S0(di + S0)}{S1(di + S1)}$$

Where,

- L1 is the work piece length,
- L0 is the blank length,
- S0 is the starting wall thickness,
- S1 is the final wall thickness
- and di is the internal diameter.

Both spinning and flow forming can also be consolidated to deliver complex parts. By rotating mandrel forming just cylindrical segments can be created. . A flat faced roller delivers a radial flange and a non orthogonal approach of nosed roller creates a bulge onward to the roller.

Features

- The one of a best features of the flow forming process allow for innovative, cost-effective engineering or redesign of your product or part, bringing about the accompanying features:
- Traditional multi-piece designs can be framed as a solitary, seamless piece.
- superior mechanical properties, for example, tensile/yield strength and hardness.
- Give design versatility to create an exceptional seamless profile with varying wall thicknesses.
- Create cylindrical, conical, or corrugated shapes up to 47" diameter.
- Typical inside finishes of 15Ra without more manufacturing steps.
- High material usage because of near-net shape forming process.

Materials Used in Flow forming

- Steels of many types e.g. stainless, carbon steel, alloy steels and heat treated steels.
- Titanium
- Brass
- Nickel
- Niobium
- Brass and Inconel
- Aluminum and Copper

Advantages

- Less costs of production.
- Wastage of material is very little.

- Surface finishes are very good.
- Components made are very accurate.
- Strength properties are good.
- High tensile strength alloys can be easily cold formed by this process. Thin walled seamless components with high precision are produced by this process.

HYDROFORMING

Hydroforming was established in the 1940's - 1950's to make available a cost effective means to create comparatively small numbers of drawn parts or parts with irregular shapes that do not afford to stamping process. Nearly all metals adept of cold forming can be hydroformed, comprising aluminum, brass, carbon and stainless steel, copper, and high strength alloys. (Koç, 2008)

Hydroforming uses a complicated shape die and a thick liquid to plastically deform the metal. This process does not uses any punch and no thinning of metal on edges and scratches result. There are two types of hydroforming, namely

- Sheet forming
- Tube forming.

The operation of a hydroforming press is similar to a female or upper die press. There is a pressurized oil chamber which is used for forming, a diaphragm of rubber and a wear pad. The male part of the die consists of a punch and ring. The punch of lower die is connected with a surrounding blank holder and a hydraulic piston. The process of hydroforming starts with setting the blank or work piece over the ring. On closing the press, the chamber of oil is brought down on to the metallic blank. The forming chamber is made pressurized with oil and the which is on the below the work piece is lifter upwards through the ring and is forced towards the oil filled forming chamber. As the female part of hydroforming press consists of a rubber chamber, the formed blank surface has sound surface finish without any cracks and scratches which are normally observed with other shaping processes. The rubber diaphragm provides support to the whole surface of metallic blank. The blank is formed into the required shape about the upward rising punch. The shape of punch is formed on the blank . on completion of hydroforming cycle the oil pressure of rubber forming chamber is released out and forming punch is withdrawn on lower side from the finished product. (Koç, 2008)

In hydroforming, controlling factor of a metal's plastic flow is the fluid pressure which is exerted on the flexible rubber diaphragm. Mostly, Fluid pressures of the order of about 100 MPa are used. The fluid pressure is homogenous and there are no sharp edges which ensure success of proper hydroforming. Many complicated shapes can be produced by employing this process. Tube hydroforming technique uses pressure of fluid to bend and pressurize the tubes. Rubber forming is extensively employed in aircraft manufacturing industry.

Tube Hydroforming

- It is employed for producing complex shapes.
- Tubes made of cold-rolled steel are used. A section of tubing is positioned inside the dies.
- Pressurized fluid is fed at both ends of the cold rolled tube.
- The tube is then hydro-formed according to the shape of the die cavity

Applications

- Automotive industry
- Sports industry
- Aircraft industry
- Household products

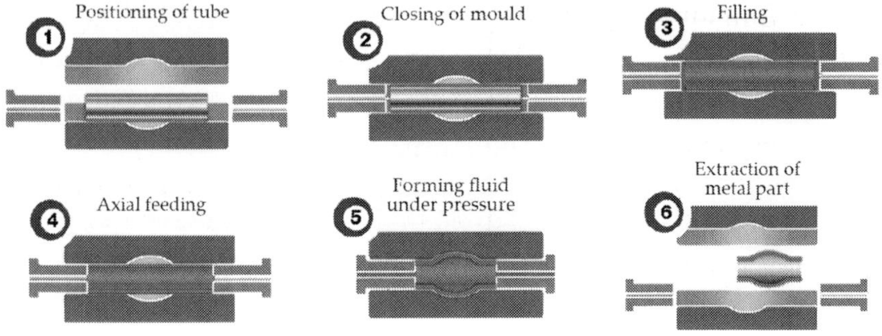

Figure 2.9 tube hydroforming process illustration (Anon., n.d.)

Sheet Hydroforming

- Sheet metal is placed into the die and water pressure is applied to force the sheet into the die cavity.
- The male punch is acting against the fluid pressure and it forms desired shape on the metallic blank

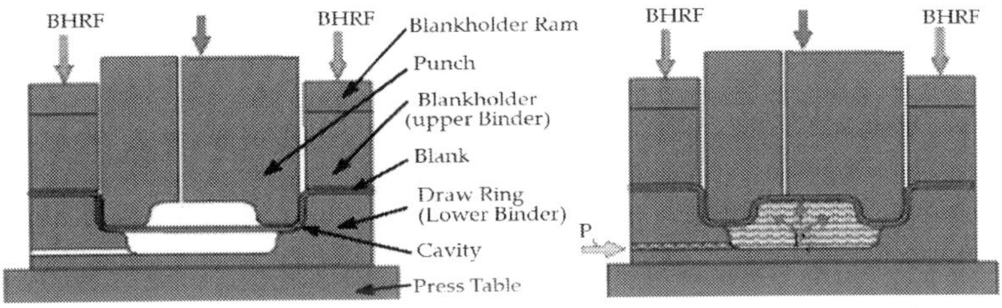

Figure 2.10 Illustration of sheet hydroforming process (Anon., n.d.)

Applications

- Automotive industry
- Sports industry
- Aircraft industry
- Household products

Advantages

- Weight of the product is reduced.
- Set-up time and tooling cost is not much higher.
- Development costs of material are low.
- Structural properties and stiffness of the product are improved.
- Less number of secondary manufacturing operations because many steps are covered in a single hydroforming process.
- Excellent dimensional control and spring back is very low.
- Many defects which are associated with matched die forming e.g. Shock lines, drawing lines, wrinkles, and tears are eliminated in hydroforming.
    - Material thinout is very low. Work-Hardening of the metal is very low with hydroforming.

- Single cycle of hydroforming can replace many steps of traditional drawing operations. Irregular shapes and contours are effectively formed. less scrap is produced. Disadvantages
- Cycle time is much longer.
- Expensive equipment is required.
- Modern welding techniques are required for joining the assembly made by hydroforming.

SINGLE CRYSTAL FABRICATION

A single crystal or mono crystalline solid is a material in which the crystal lattice of the whole substance is consistent and unbroken up to all the surfaces of the specimen, with no grain boundaries. The nonappearance of the imperfections associated with grain boundaries can give mono crystals novel properties, especially mechanical, optical and electrical, which can likewise be anisotropic, contingent upon the kind of crystallographic structure. These properties, along with making them valuable materials, are commercially utilized as a part of innovative technology applications, particularly in optics and gadgets manufacturing.

Since entropic impacts support the existence of a few defects in the microstructure of solids, such as presence of other elements, inhomogeneous strain and crystallographic deformities such as dislocations, consummate single crystals of important size are exceedingly uncommon in nature, and are additionally hard to deliver in the lab, however they can be made under controlled conditions. Then again, single crystals having imperfections can achieve gigantic sizes in nature: a few mineral species such as beryl, gypsum and feldspars are known to have single crystals a few meters over.

On account of silicon and metal single crystal creation the procedures utilized include very controlled and in this way moderately slow crystallization. Particular processes to deliver extensive single crystals incorporate the Czochralski procedure and the Bridgman method. Different less intriguing strategies for crystallization might be utilized, contingent upon the physical properties of the substance, including sublimation, hydro-thermal synthesis or just solvent based crystallization. (Hans J. Scheel, n.d.) An alternate route to making single crystalline materials is called epitaxy. As of 2009, this procedure is utilized to store thin (micrometer to nanometer scale) layers of the same or distinctive materials on the surface of a current single crystal. Utilizations of this method lie in the areas of semiconductor production, with potential uses in other nanotechnological fields and catalysis.

Czochralski Process

The Czochralski procedure is a process for crystal development used to get single crystals of semiconductors, metals, salts and manufactured gemstones. The process is named after Polish researcher Jan Czochralski, who designed the process in 1915 while examining the crystallization rates of metals. He made this disclosure coincidentally, while investigating the crystallization rate of metals, rather than dunking his pen into the ink, he did as such in liquid tin and drew a tin fiber, that later ended up being a single crystal. (Hans J. Scheel, n.d.)

The most imperative application might be the development of extensive round and hollow ingots, or boules, of single crystal silicon utilized as a part of the gadgets business to make semiconductor gadgets like integrated circuits. Different semiconductors, such as gallium arsenide, can likewise be developed by this process, despite the fact that lower deformity densities in this case can be utilizing variations of the Bridgman-Stockbarger process.

Crystal sizes

Because of the efficiencies of regular wafers, the semiconductor business has utilized wafers with standardized measurements. In the past days, the boules were littler, just a couple of inches wide. With cutting edge technology innovation, top of the line gadget makers utilize 200 mm and 300 mm measurement wafers. The width is controlled by exact control of the temperature, the rates of revolution and the speed with which the seed holder is pulled back. The crystal ingots from which these wafers are cut can be up to 2 meters long, measuring a few hundred kilograms. Bigger wafers permit enhancements in manufacturing productivity, as more chips can be created on every wafer, so there has been a relentless drive to increase silicon wafer sizes. The subsequent stage up, 450 mm, is as of now booked for presentation in 2018.[Silicon wafers are regularly around 0.2–0.75 mm thick, and can be cleaned to extraordinary levelness for making integrated circuits or finished for making sunlight based cells.

(Hans J. Scheel, n.d.) The procedure starts when the chamber is warmed to roughly 1500 degrees Celsius, liquefying the silicon. At the point when the silicon is completely liquefied, a little seed crystal mounted on the end of a pivoting shaft is gradually brought down until it just plunges beneath the surface of the liquid silicon. The pole turns counterclockwise and the crucible pivots clockwise. The turning shaft is then drawn upwards gradually—around 25 mm for each hour when making a crystal of ruby, enabling a generally barrel shaped boule to be produced. The boule can be from one to two meters, contingent upon the measure of silicon in the crucible.

The electrical attributes of the silicon are controlled by including material like phosphorus or boron to the silicon before it is softened. The additional material is called dopant and the procedure is called doping. This technique is additionally utilized with semiconductor materials other than silicon, such as gallium arsenide.

Incorporating Impurities

At the point when silicon is developed by the Czochralski technique, the liquefy is contained in a quartz cruicible. Amid development, the inside of the crucible break up into the melted silicon and Czochralski silicon in this manner contains oxygen at an ordinary range of 1018 cm−3 . Oxygen content can have advantageous or hindering impacts. intelligently picked tempering / annealing conditions can offer ascent to the arrangement of oxygen atoms. These have the impact of catching undesirable metal contaminations in a procedure known as gettering, enhancing the quality of encompassing silicon. otherwise, forming of oxygen precipitate at unintended areas can likewise wreck electrical structures. (Hans J. Scheel, n.d.) Furthermore, oxygen impurities can enhance the mechanical quality of silicon wafers by immobilizing any disengagements which might be presented amid gadget preparing. It was tentatively presented in the 1990s that the high oxygen fixation is additionally advantageous for the radiation hardness of silicon molecule locators utilized as a part of hard radiation environment. Therefore, radiation indicators made of Czochralski-and Magnetic Czochralski-silicon are thought to be promising candidates for some future high-energy material science experiments. It has likewise been demonstrated that the content of oxygen in silicon increases catching impurities amid post-implantation strengthening processes. Oxygen content can respond with boron in a lit up environment, such as that accomplished by sun oriented cells. This outcomes in the making of an electrically dynamic boron–oxygen complex that takes away from cell execution. Module yield drops by roughly 3% amid the initial couple of hours of light presentation.

Bridgman–Stockbarger Technique

The process of Bridgman–Stockbarger is named after Harvard physicist Percy Williams Bridgman and MIT physicist Donald C. Stockbarger. The procedure incorporates two comparative however unmistakable techniques fundamentally utilized for developing boules (single crystal ingots), yet which can be utilized for cementing polycrystalline ingots as well. (Hans J. Scheel, n.d.)

The techniques include heating polycrystalline material over its dissolving point and gradually cooling it from one end of its holder, where a seed crystal is found. A single crystal of an known crystallographic direction from the seed material is developed on the seed and is continuously framed along the length of the compartment. The procedure can be completed in a horizontally leveled or vertical direction, and as a rule includes a pivoting pot to mix the melt.

The Bridgman process is a famous method for creating certain semiconductor crystals such as gallium arsenide, for which the Czochralski procedure is more troublesome. The procedure can dependably deliver single crystal ingots, however does not really bring about uniform properties through the crystal.

The contrast between the Bridgman procedure and Stockbarger procedure is inconspicuous: While the two strategies use a temperature difference and a moving pot, the Bridgman system uses the generally uncontrolled inclination created at the exit of the heater; the Stockbarger method presents a puzzle, isolating two coupled heaters with temperatures above and beneath the point of solidification. Stockbarger's alteration of the Bridgman method takes into consideration better control over the temperature difference at the liquid/crystal interface. (Hans J. Scheel, n.d.) At the point when seed crystals are not utilized as portrayed above, polycrystalline ingots can be created from a feedstock comprising of rods, lumps, or any sporadically formed pieces once they are liquefied and permitted to reset. The resultant microstructures of the ingots so acquired are normal for directionally solidified metals and alloys with their aligned grains.

Hydrothermal Synthesis

Hydrothermal synthesis incorporates the different strategies of crystallizing substances from high temperature fluid arrangements at high vapor pressures, additionally named "Hydrothermal process"

Hydrothermal synthesis can be characterized as a strategy for union of single crystals that relies upon the dissolvability of minerals in hot water under high pressure. The crystal development is performed in a mechanical assembly comprising of a steel made pressure vessel called an autoclave, in which a supplement is provided alongside water. A temperature difference is kept up between the opposite closures of the development chamber. At the hotter end the supplement solute disintegrates, while at the cooler end it is kept on a seed crystal, developing the required crystal. (Hans J. Scheel, n.d.)

Preferences of the Hydrothermal synthesis over different sorts of crystal development incorporate the capacity to make crystalline phases which are not stable at the softening temperature. Additionally, materials which have a high vapor pressure close to their dissolving points can be developed by the Hydrothermal synthesis. The technique is additionally, especially reasonable for the development of vast great quality crystals while keeping up control over their organization. Burdens of the technique incorporate the need of costly autoclaves, and the inability of watching the crystal as it develops.

Methods

Temperature-Difference Method

This is the most broadly utilized technique in Hydrothermal synthesis and crystal development. Super saturation is accomplished by decreasing the temperature in the crystal development zone. The supplement is set in the lower some portion of the autoclave loaded with a particular measure of solvent. The autoclave is warmed so as to make two temperature zones.

The supplement disintegrates in the hotter zone and the saturated aqueous solution in the lower part is transported to the upper part by convective movement of the solution. The cooler and denser arrangement in the upper piece of the autoclave drops while the counter flow of solution ascends. The solution ends up noticeably supersaturated in the upper part as the consequence of the decrease in temperature and crystallization sets in.

Temperature-Reduction Technique

In this method crystallization happens without a temperature difference between the development and disintegration zones. The super saturation is accomplished by a continuous decrease in temperature of the solution in the autoclave. The disservice of this method is the trouble in controlling the development procedure and incorporating seed crystals. Hence, this strategy is sometimes utilized.

Metastable-Phase Technique

This method is based on the distinction in solubility between the phase to be developed and that acting in as the beginning material. The supplement comprises of compounds that are thermodynamically not stable under the development conditions. The dissolvability of the metastable phase surpasses that of the steady phase, and the last crystallize because of the disintegration of the metastable phase. This method is normally consolidated with one of the other two methods above.

Uses

A substantial number of compounds having a place in all classes have been developed under hydrothermal synthesis conditions: pure elements, simple oxides,complex oxides, molybdates, germanates etc. hydrothermal synthesis is ordinarily used to develop manufactured quartz, jewels and other single crystals with business esteem. A portion of the crystals that have been effectively developed are emeralds, rubies, quartz and others. The technique has turned out to be to a great degree productive both in the research for new compounds with particular physical properties and in the systematic physicochemical examination of specific multicomponent systems at raised temperatures and pressures. (Hans J. Scheel, n.d.)

Epitaxy

Epitaxy implies to the developing a crystalline overlayer on a crystalline material substrate.

The overlayer is known as epitaxial layer. The term epitaxy originates from the Greek roots epi , implying "above", and taxis, implying"a we-arranged fashion". For most innovative applications, it is needed that the material which is deposited as a crystalline overlayer has one all around characterized direction with respect to the substrate crystal structure.

Epitaxial layers might be developed from gaseous or fluid antecedents. Since the substrate goes about as a seed crystal, the deposited film may bolt into at least one crystallographic orientation with regard to the substrate crystal. if the overlayer either shapes a random orientation with respect to the substrate or does not produce a well-arranged overlayer, it is named non-epitaxial development. On the off chance that an epitaxial film is deposited on a substrate of a similar composition, the procedure is called homoepitaxy; else it is called heteroepitaxy.

Homoepitaxy

It is a sort of epitaxy performed with just a single material, in which a crystalline film is developed on a substrate or film of a similar material. This innovation is utilized to grow a film which is more immaculate than the substrate and to manufacture layers having distinctive doping levels. In scholarly writing, homoepitaxy is frequently shortened to "homoepi". (Hans J. Scheel, n.d.)

Heteroepitaxy

It is a sort of epitaxy performed with materials that are not the same. In heteroepitaxy, a crystalline film develops on a crystalline substrate or film of an alternate material. This innovation is frequently used to develop crystalline layers of materials for which crystals can›t generally be gotten and to create incorporated crystalline layers of various materials. Examples comprise of silicon on sapphire, gallium nitride (GaN) on sapphire, AlGaInP on GaAs or precious stone or iridium and graphene on hexagonal boron nitride. (Hans J. Scheel, n.d.)

Heterotopotaxy

It is a procedure like heteroepitaxy with the exception that thin film development is not restricted to two-dimensional development; the substrate is comparative just in structure to the thin-film material.

Pendeo-Epitaxy

It is a procedure in which the heteroepitaxial film is developing vertically and along the side in the meantime. In 2D crystal heterostructure, graphenen nanoribbons implanted in hexagonal boron nitride are a case for pendeo-epitaxy. Epitaxy is utilized as a part of silicon-based manufacturing processes for bipolar junction transistors and present day correlative metal–oxide–semiconductors, yet it is especially critical for compound semiconductors such as gallium arsenide. Manufacturing issues comprise of control and consistency of the deposited layer's resistivity and thickness, the cleanliness and virtue of the surface and the chamber enviornment, the counteractive action of the commonly substantially more profoundly doped substrate wafer's dissemination of dopant to the new layers, defects of the development procedure, and ensuring the surfaces amid the fabrication and handling. (Hans J. Scheel, n.d.)

Methods

Epitaxial silicon is generally developed utilizing vapor-phase epitaxy (VPE), an alteration of chemical vapor deposition.Molecular-beam and liquid-phase epitaxy (MBE and LPE) are likewise utilized, for the most part for compound semiconductors. Solid phase epitaxy is utilized essentially for recovering the crystal-damage. (Hans J. Scheel, n.d.)

Vapor-Phase

Silicon is most ordinarily deposited by doping with silicon tetrachloride and hydrogen at around 1200 to 1250 °C

$$SiC14(g) + 2H2(g) \leftrightarrow Si(s) + 4HCl(g)$$

This is a reversible reaction, and the development rate depends firmly upon the amount of the two source gasses. Development rates over 2 micrometers for every minute deliver polycrystalline silicon, and negative development rates (etching) may happen if an excessive amount of hydrogen chloride byproduct is available. (sometimes, hydrogen chloride might be added deliberately to etch the wafer.) An extra etching response contends with the deposition reaction:

$SiC14(g) + Si(s) \leftrightarrow 2SiC12(g)$

Silicon VPE may likewise utilize silane, dichlorosilane, and trichlorosilane source gasses. For example, the silane reaction happens at 650 °C along these lines:

$SiH4 \rightarrow Si + 2H2$

This reaction does not accidentally etch the wafer, and happens at less temperatures than deposition from silicon tetrachloride. In any case, it will shape a polycrystalline film unless firmly controlled, and it permits oxidizing species that hole into the reactor to taint the epitaxial layer with undesirable mixes such as silicon dioxide.

VPE is classified by the science of the source gasses, such as hydride VPE and metalorganic VPE. (Hans J. Scheel, n.d.)

Liquid-Phase

Liquid phase epitaxy (LPE) is a method to develop semiconductor crystal layers from the liquid on solid substrates. This occurs at temperatures well underneath the dissolving point of the semiconductor which is deposited. The semiconductor is broken down in the solvent of another material. At conditions that are near the harmony amongst disintegration and deposition, the deposition of the semiconductor crystal on the substrate is moderately fast and uniform. The most utilized substrate is indium phosphide (InP). Different substrates like glass or earthenware can be used for many applications. To encourage nucleation, and to maintain no tension in the developed layer the heat expansion coefficient of substrate and developed layer ought to be comparative. (Hans J. Scheel, n.d.)

Solid-phase

Solid Phase Epitaxy (SPE) is a move between the indistinct and crystalline phases of a material. It is normally done by first storing a deposition of amorphous material on a crystalline substrate. The substrate is then warmed to crystallize the film. The single crystal substrate fills in as a format for crystal development. The annealing step used to recrystallize or mend silicon layers amorphized amid particle implantation is likewise viewed as one kind of Solid Phase Epitaxy. The Impurity isolation and redistribution at the developing crystal-nebulous layer interface amid this procedure is utilized to consolidate low-dissolvability dopants in metals and Silicon.

Molecular-Beam Epitaxy

In molecular beam epitaxy (MBE), a source material is warmed to deliver a vanished light emission. These particles go through a high vacuum (10−8 Pa; for all intents and purposes free space) to the substrate, where they consolidate. MBE has decreasing throughput than different types of epitaxy. This method is broadly utilized for developing III, IV, and V group semiconductor crystals.

Doping

An epitaxial layer can be doped amid deposition by adding debasements to the source gas, such as arsine, phosphine or diborane. The centralization of impurity influence in the gas phase decides its fixation in the deposited film. As in chemical vapor deposition (CVD), impurities alter the testimony rate. Moreover, the high temperatures at which CVD is performed may permit dopants to diffuse into the developing layer from different layers in the wafer. Additionally, dopants in the source gas, freed by vanishing or wet etching of the surface, may diffuse into the epitaxial layer. The dopant profiles of basic layers change as well, however not as essentially.

Applications

Epitaxy is utilized as a part of nanotechnology and in semiconductor creation. In reality, epitaxy is the main moderate strategy for brilliant crystal development for some semiconductor materials. In surface science, epitaxy is utilized to make and study monolayer and multilayer movies of adsorbed natural atoms on single crystalline surfaces. Adsorbed particles ordered structures on molecularly level porches of single crystalline surfaces and can straightforwardly be watched by means of examining burrowing microscopy. In contrast, surface deformities and their geometry have noteworthy impact on the adsorption of natural molecules (Hans J. Scheel, n.d.)

CHAPTER 3

Welding and Casting Processes

INTRODUCTION

Present day manufacturing incorporates every single procedure required in the production and reconciliation of a product's parts. Joining is an essential procedure in various industries, such as aviation, car, oil, and gas. Numerous products can't be manufactured as a solitary piece, so parts are created first and assembled later. Joining processes can be classified as a liquid-solid-state process and mechanical means. Liquid-solid-state processes incorporates welding, brazing, binding, and cement holding. Mechanical joining incorporates fasteners, fasteners, nuts, and screws.

Mostly metal joining by welding is a procedure that utilizes warmth to liquefy or warm metal just underneath the softening temperature. Joining metal by heating and softening is known as fusion welding. Without fusion, the procedure is known as solid state welding. fusion welding incorporates arc welding and laser welding, Whereas solid state welding such as friction stir welding (FSW) where joining happened underneath the softening temperature.

Casting is a manufacturing procedure in which a fluid material is typically filled in a mold, which contains an empty depression of the coveted shape, and then permitted to set. The set part is otherwise called a casting, which is shot out or broken out of the shape to finish the procedure. Casting materials are typically metals or different cold setting materials that cure subsequent to combining at least two parts; its illustrations are epoxy, solid, plaster and mud. Casting is frequently utilized for making complex shapes that would be generally troublesome or uneconomical to make by different strategies.

FRICTION STIR WELDING

Wayne Thomas developed Friction Stir Welding (FSW) was by at TWI Institute in 1991. At start, the FSW process was confined to laboratory works but in a short time it was established that FSW has many outstanding advantages in the welding of aluminum alloy products. Friction Stir Welding is a joining process which bonds metals in solid-state, which implies that the materials are joined with each other without heating them to their melting point. This is a breakthrough in the welding technology. By Use of FSW, high speed accurate and excellent quality welding of aluminum alloys of 7xxx and 2xxx series are now made possible which were conventionally regarded as unweldable alloys. (Daniela Lohwasser, 2009)

Friction stir welding (FSW), exemplified in Figure. 1, is a solid state bonding procedure which uses a rotating tool that travels laterally with the line of joint amongst two workpieces, which are to be joined and generates extensive heat due to friction. This tool also mechanically stirs the metal for producing the seam of welding. The FSW welding process is named after the action of rotating tool i.e. stirring and mixing actions. FSW is different from traditional Resistance welding processes because heat for fusion is generated by the hard friction tool while in resistance welding this heat is generated by friction of work pieces with each other. (Company, 2004)

The rotating tool used for FSW has stepped geometry. It is comprised of a probe and a cylindrical shoulder. The probe is projected at the bottom of cylindrical shoulder. In the course of welding, the cylindrical shoulder moves at high speed over the upper surfaces of the two work pieces which are to be joined. Rubbing action of the shoulder produce friction heat for joining the metals. The geometry of probe is specially developed to produce mixing action Probe serves as a mixing tool by generating additional heat along the joint surface of work pieces. The heat generated by both the friction and mixing actions of friction tool does not bring the metal in liquid state but it softens the metal to a very fused and soft condition. (Daniela Lohwasser, 2009)

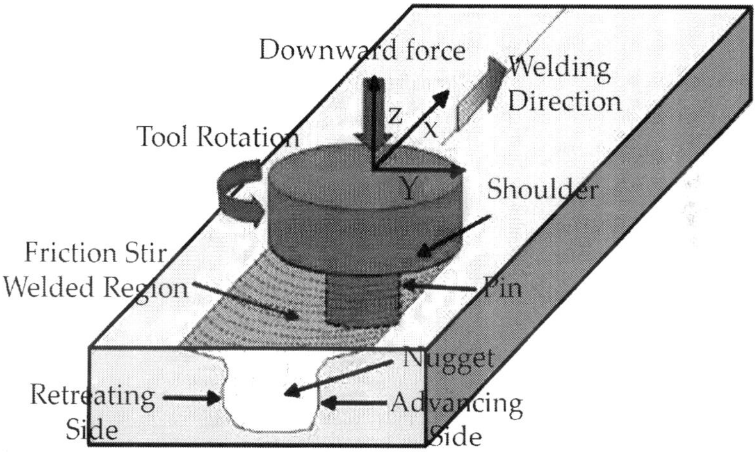

Figure 3.1. Illustration of Friction stir welding (FSW) (Anon., n.d.)

By onward motion of friction tool along the length of joint, the front side of the rotating probe pushes the metal and develops a forging action into the seam of weld joint. The shoulder functions to contain the soft fused metal in the joint seam area around the rotating probe.

Friction Stir Welding is employed for joining aluminium sheets without aid of any filler wire or any shielding gas. Metal sheet thicknesses of 0.5 to 65 mm are easily welded by using FSW technique without producing any defects such as porosity and cracking. Initially friction stir welding was used for non-ferrous alloys, but today, friction stir welding is also been used for welding a wide variety of materials.

To guarantee high reproducible quality with Friction stir welding, the welding equipment must be made more automatic. Traditional CNC setups can easily produce simple welds, but to produce welds of greater thicknesses, more sophisticated equipment is required.

Process Characteristics

The Friction stir welding process is a unique welding process in which joint is formed at temperatures which are lower than base metal's melting temperature. (Jr, 2011) The heat produced in the area of joint is normally in the range of 80-90% of that produced at the melting temperature. For making production welds by using a welding technique, welding procedure specification and procedure qualification record documents are essentially required. Heat input to weld area is of major concern in preparing these documents. In arc welding, current and voltage are the major players in heat input formula. In case of Friction stir welding, the heat input is solely based on mechanical parameters such as frictional force, rotation and stirring. Many investigations have been made for studying the heat generation and propagation in the weld area in friction stir welding. A generalized model is illustrated in the following equation:

$Q = \mu\omega FK$

in this equation,

- Q is the heat generated from friction
- μ is the speed of rotation of the tool
- F Is the down force acting on the weld area
- K is the geometry constant of the tool

The properties of Friction stir welded joint components are mostly superior than conventionally fusion-welded components. The properties of major advantage include, fatigue and toughness properties.

Welding Parameters

In giving legitimate contact and in this manner guaranteeing a brilliant weld, the most vital control factor is downward drive force. This ensures high caliber welds even where size tolerance problems in the materials to be joined may emerge. It likewise empowers strong control at higher welding speeds, as the down force will guarantee the production of frictional heat for making the material soft for fusion. When utilizing FSW, the accompanying parameters must be controlled: down compel, welding speed, the turn speed of the welding device and tilting edge. Just four fundamental parameters need to be aced, making FSW perfect for automated welding

Advantages

- Superior mechanical characteristics of the welded joints.
- Friction stir welding eliminates Toxic fumes, warping, shielding problems, and many more issues related to arc welding.
- Distortion and cooling shrinkage of the weld is low.
- Weld is aesthetically attractive.
- Almost no or little post-weld treatment is required.
- Friction stir welding is enviorment friendly because no shielding and toxic materials are used. impact on the environment
- Friction stir welding is an energy saving process. Weld-joint area does not needs much preparation.
- Joint efficiency of friction stir welds is high.
- The process is energy efficient
- Heat input to material is low. Fatigue life of the component is high.

Disadvantages

- On withdrawing the stir welding tool from the work piece, an exit hole is always generated.
- Very firm gripping of the workpieces is required and heavy-duty clamps are used for this purpose.
- Large amount of force is required for welding.

Application

Friction stir welding is used chiefly for welding aluminum alloys of small to large sizes. Steel, titanium, copper, plastics and composite materials are also friction stir welded.

FSW is employed in

- Aerospace,
- Automotive
- Shipbuilding industries.

PLASMA KEYHOLE WELDING

The principle of plasma keyhole welding is same as the principle of plasma welding. A transferred arc operating system is employed for plasma keyhole welding, the plasma gas stream and current are generally expanded and the opening size might be decreased. The correct current where the keyhole mode is started will rely on upon the torch geometry and the joint material and thickness, yet currents of more than 200 A and plasma gas streams of 3 to 4 1 min–1 are run of the mill with a 2 to 3 mm measurement narrow opening. Under these conditions high arc pressures are created by electromagnetic narrow path. (K. G. Swift, 2013)

The thermal productivity of the procedure is high and it has been assessed that the warmth exchanged to the workpiece from a 10 kW plasma arc can be as high as 66% of the aggregate procedure power

Control of Plasma Keyhole Welding

The parametric connections for plasma keyhole welding are complex by varieties because of torch geometry and it is likely that parameters created utilizing a particular torch won't be transferable to a torch of an alternate design. The control factors might be grouped, as presented in Table, into those ordinarily used to coordinate the conditions to the application, the primary controls, and those regularly picked and settled before alteration of the procedure, the secondary components. (Company, 2004)

The impact of the principle control parameters on control might be stated as follows.

Mean Welding Current

The arc pressure delivered by magnetic narrow path is related to the square of the mean current. With high currents, high arc strengths are produced furthermore, undercut or bumping may happen. The current must be controlled in conjunction with speed of welding to create the required welding seam profile. (K. G. Swift, 2013)

Table 3.1 Plasma key hole welding controlling areas (Norrish, n.d.)

Primary	Secondary
Current	Orifice shape/type
Travel speed	Orifice diameter
Plasma gas flow	Electrode vertex angle Electrode set back Shielding and plasma gas type Torch stand off

Plasma Gas Flow Rate

The existence in the opening of an insulating layer of non-ionized gas encompassing the plasma is necessary. This insulating layer may separate at exorbitant currents, at low plasma gas streams or when lack of cooling is there. The resultant weaknesses will regularly cause twofold arcing and much harm to the torch. The plasma gas stream rate additionally impacts the arc pressure and, on the off chance that it is too low, the keyhole impact might be lost. At high currents, it is in this way important to keep up the plasma gas stream rate at a sensibly normal state, however as far as possible will be dictated by the event of undercut and reduced thermal proficiency. (R.L. Timings, 2009)

Welding Speed

In the event that the speed is too high undercutting and fragmented penetration will come about, though at low speeds the keyhole size may wind up plainly over the top and the weld pool will crumple. The working scope of this procedure is along these lines decided by a blend of mean current, welding pace and plasma gas stream. The connection between thickness entered and welding speed at different plasma power levels is shown in Figure.

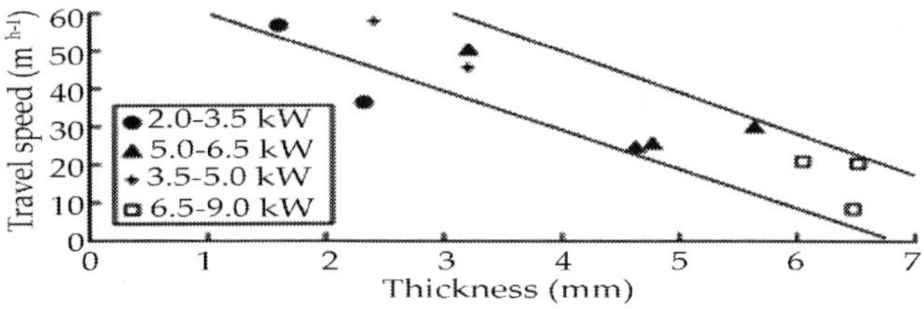

8.3 Thickness versus travel speed for plasma keyhole welding of stainless steel at various power levels. (Data taken from a range of studies.)

Figure 3.2 Graph of plasma key hole welding for stainless steel at different powers (Norrish, n.d.)

Secondary Controls

Orifice Diameter.

In the event that the speed is too high undercutting and fragmented penetration will come about, though at low speeds the keyhole size may wind up plainly over the top and the weld pool will crumple. The working scope of this procedure is along these lines decided by a blend of mean current, welding pace and plasma gas stream. The connection between thickness entered and welding speed at different plasma power levels is shown in Figure. Orifice diameter. Diminishing the orifice measure builds the arc pressure and voltage. For keyhole welding, 2–3 mm diameter, orifices are ordinarily utilized.

Electrode Geometry

The electrode geometry and its position inside the torch are basically required because of their impact on gas stream inside the torch. It is proposed that tolerances of 0.1 to 0.2 mm on electrode position are required. Concentricity of the electrode is likewise vital, since any misalignment may result in uneven arc conduct and poor weld seam appearance; the electrode ought to either be customizable or settled by a fired embed inside the torch.

Multiport Nozzles

The multiport spout might be utilized to upgrade narrowing furthermore, delivering an elliptical arc profile which is extended along the pivot of the weld. A moment concentric spout may additionally be utilized to give an expansion in arc weight, albeit exorbitant centering gas stream rates may diminish the thermal effectiveness of the procedure.

Shielding Gasses.

The most widely recognized shielding and plasma gas is argon. From 1 to 5% hydrogen might be added to the argon shielding gas for welding low-carbon and stainless steels of austenitic type.3 The impact of these little increments of hydrogen is very huge, giving enhanced weld seam cleanliness, higher travel speeds and enhanced narrowing of the arc. Helium might be utilized as a shielding medium for high-conductivity metals. It will tend to expand the aggregate warmth input in spite of the fact that it might lessen the impact of the narrow opening and deliver a more diffused heat source. With 30% helium/70% argon shielding gas blends, keyhole welding speeds 66% higher than those accomplished with argon shielding have been accounted for aluminum. The shielding gas stream is not regular restriction in spite of the fact that adequate gas ought to be given to deliver successful shielding of the weld region.

Arc Characteristics

Narrowing of any arc causes an upsurge in arc voltage. In the plasma control, expanding the plasma gas stream, diminishing the diameter of the constriction, upsurging current and adding hydrogen or helium to the gas will all upsurge the arc voltage. The impact of narrowing is presented in figure. The thermal profile of plasma arcs typically takes after a Gaussian distribution as presented in Figure

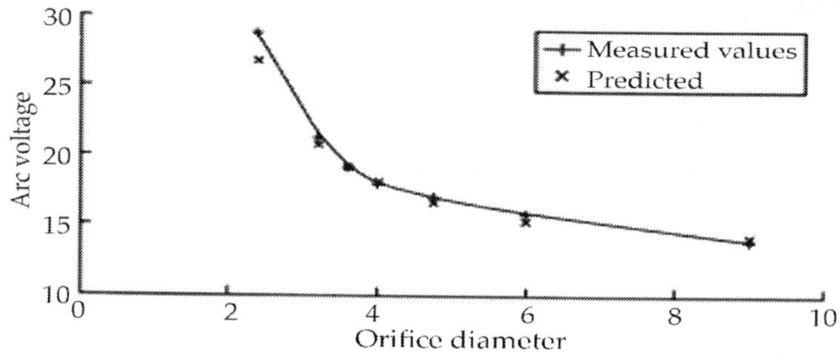

Figure 3.3 Effect of orifice size on welding voltage (Norrish, n.d.)

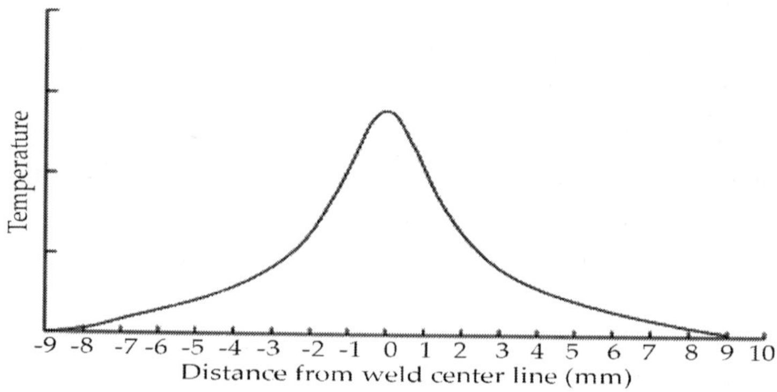

Figure 3.4 Temperature profile on work piece material (Norrish, n.d.)

Reverse Polarity Plasma

The plasma procedure is typically worked with the electrode negative (DCEN) also, the workpiece positive. This polarity might be reversed to permit cathodic cleaning to happen when welding aluminum amalgams. It is generally essential to upsurge the electrode size and lessen the most extreme current due to the extra warming impact inside the torch, yet helium shielding gas might be used to upsurge the thickness go weldable in the keyhole mode up to 8 mm.

Variable Polarity Plasma

The consolidated advantages of DCEN and DCEP operation might be accomplished by utilizing a variable polarity control supply. Astounding outcomes have been acquired in welding aluminum alloys with this arrangement.

Pulsed Keyhole Plasma

The typical keyhole method of operation is confined to welding in the downhand position and requires exceptionally basic control of speed as examined previously. The tolerance of the procedure might be enhanced by regulating the current and positional welds might be made in an extensive variety of metals and plate thicknesses. The current is regulated at moderately low frequencies, the beat time and amplitude being founded on the necessity to set up a keyhole. but, the necessary conditions being set to keep up an arc however permit solidifying. The resultant weld is consequently shaped from a progression of overlaping detects, the travel speed being changed in accordance with gives no less than 60% overlapping. Pulsed operation enhances the imperviousness to undercut and by and large produces a more extensive, compliment seam. Cases of run of the mill applications and the components of the DC and pulsed methods of operation are presented underneath.

Applications

The keyhole method of operation makes it conceivable to perform single-pass square butt welds from one side in plate thicknesses up to around 10 mm.

Carbon–Manganese Ferritic Steel.

The plasma keyhole welding of carbon– manganese steels has as of late been assessed for circumferential root runs in pipes for power production and seashore applications. The utilization of the controls empowers thick root segments (6–8 mm) to be welded in a solitary pass from one side and altogether enhances profitability. Pulsed plasma keyhole has been utilized for these examinations to enhance working tolerance and positional execution.

Austenitic Stainless Steel.

Austenitic stainless steel might be promptly welded utilizing the keyhole method and the procedure has been connected to the longitudinal welding of pipe and also the fabrication of segments for cryogenic benefit. Welding speeds of around 1.0 m min–1 are achievable with keyhole welds in material up to 2.7 mm thick, while welds in 6.0 mm thick material might be made at 0.35 mmin–1. The utilization of hydrogen in the shielding gas or exclusive blends containing from 1 to 5% hydrogen give enhanced seam appearance and improved travel speed. Undercut might be constrained via watchful control of welding parameters, in any case, if this is most certainly not conceivable, pulsed operation or the utilization of restorative passes is suggested.

SQUEEZE CASTING

Gas holes and shrinkage porosity are major defects which limit the vitality of casting processes and performance of cast components. A number of efforts have effectively extended the competences of conservative die casting processes without compromising the economic benefits. Squeeze casting is developed to remove these defects by using the two strategies mentioned below:

- Removing or decreasing the gases which are trapped in the molten metal.
- Removing or decreasing the plastic aging. (K. G. Swift, 2013)

Squeeze casting is a Blend of casting and forging operations. It casts the molten metal by pouring into a preheated lower die, it is covered with upper die to provide a mold cavity for solidification of the metal. Squeeze casting differs from the conventional die casting processes in which the die halves are clamped with each other before injecting the metal into it. Because of the mixed type of the process, it is also termed as liquid metal forging process. (R.L. Timings, 2009) Squeeze casting or liquid-metal forging, is a metal casting and forging process in which molten metal is poured in the closed die placed between the stages of a powerful hydraulic press. The high pressure applied by the press and good thermal contact of the molten metal with the metallic die result in a steep thermal gradient which causes rapid heat transfer from metal to die and produces a fine casting product which is free of pores and has small-grain structure. Squeeze casting products have superior mechanical properties similar to forged or rolled products. The squeeze casting process can be conveniently mechanized for producing precise and near-net shaped products with excellent quality. (Benedict, 1987)

The process of squeeze casting was established in 1960 in USA and it has been widely used for casting nonferrous products. Aluminum alloys, copper alloys and magnesium alloys are easily casted by using this technique. A number of ferrous alloy products having comparatively simple geometry have been produced by using squeeze casting process.

The squeeze casting process blends the benefits of both casting and forging operations. Application of high pressure during solidification, eliminates all the porosities caused by gas and solidification shrinkage. It also increases the rate of cooling of the casting by increasing the contact between die surface and cast metal. High pressure, increased contact with die surface and high cooling rate all combine to form a fine grained casting with superior properties.

There is a major disadvantage of squeeze casting which is known as Macro segregation. Macro segregation happens due to lack of flow of fluid between the dendrites. As the cooling rate is fast and metal is in pressurized condition, the density differences arise, solidification contractions are disturbed and macro segregation is the obvious result. Macro segregation is most easily observed with aluminum alloys. Macro segregation results in non-uniform macrostructure and poor mechanical properties. Application of pressure during solidification effectively minimizes the formation of shrinkage and gas porosities but it can produce macro segregation. By carefully controlling all the parameters of squeeze casting macro segregation can be effectively eliminated for producing sound castings. (Company, 2004)

Squeeze casting is not complicated and economically justified process with good process efficiency and quality products. It can be readily shifted to automatic process with high production rates and reproducible quality. The casting products manufactured by using squeeze casting process exhibit maximum mechanical characteristics which can be produced with a casting process. The critical applications of squeeze casting products sometimes required close control over solidification rate, pressure and refined microstructure which can be easily attained with good shop practice.

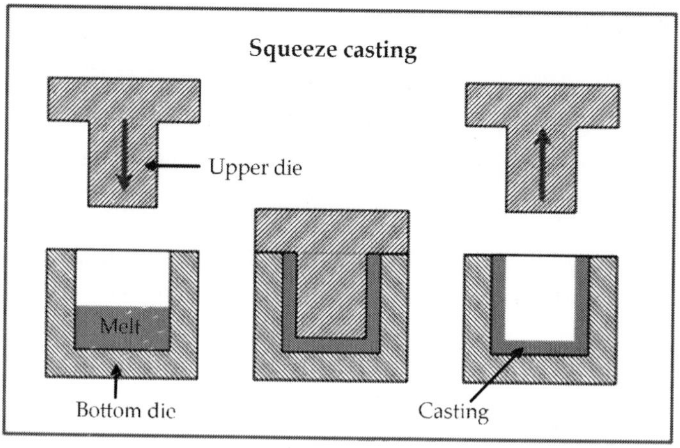

Figure 3.5 Illustration of Squeeze casting process (Anon., n.d.)

Fig. illustrates that, squeeze casting comprises of entering liquid metal into a preheated, lubricated die and metal is forged while it sets. The pressure is connected not long after the metal starts to solidify and is kept up until the point that the whole casting has set. Casting discharge and its management in die are done similarly as in close die forging. There are various factors that are for the most part controlled for the quality and nature of the castings. (Benedict, 1987)

Casting Parameters

Casting temperatures is determined by the type of alloy which is being cast and the geometry. The starting temperature for squeeze casting is generally in the range of 6to 55°C higher than the liquidus temperature of the alloy. Temperatures of the dies and tools vary from 190 to 315°C for frequent use. The time duration between two events of pouring of hot metal and the application of pressure for collapsing the early solidifying webs in the hot metal is known as time delay. Pressure levels ranging from 50 to 140 MPa are generally employed in squeeze casting. The time for which pressure is kept applied at the dies ranges from 30 to 120 seconds for a casting weight of 9 kilograms. Lubricants are also used in squeeze casting, normally graphite powders and graphite colloid sprays are used for squeeze casting of non-ferrous alloys. (K. G. Swift, 2013)

Advantages

- Many complicated and complex shapes which are not producible from other casting processes can be easily produced by squeeze casting process.
- Less or no post casting steps are required
- Porosities in the cast product are very less.
- The surface finish of the products is very good.
- The micro structure of the casting is homogenous and fine which results in superior mechanical properties.
- Wastage of material is very low because no post-cast machining is required.
- No chances for gas entrapment.
- The casting produced can be heat treated.

Limitations

- Dies and tools are sophisticated and expensive.
- The process tooling which is used for one type of materials is not suitable for other type of materials.
- The precise process control is required for producing accurate products of repeated quality and low cycle time.
- Cost of equipment is high.
- High production costs means this process is suitable for only large volume production.

Application

- Aerospace,
- Automotive
- Shipbuilding industries.

HYBRID EVAPORATIVE PATTERN CASTING PROCESS

Hybrid Evaporative Pattern Casting Process is a variant of evaporative pattern casting processes in which pattern is made up of an evaporative material. Pattern is coated with a refractory material and is placed in a mold. A riser is provided in the mold cavity so that vapors of the pattern can escape through it. Subsequently, the mold is heated to evaporate the pattern material. The pattern material evaporates off and creates a mold cavity in which molten metal is poured. (K. G. Swift, 2013)

One of the confinements of evaporative pattern casting process is that the sand, which is free flowing, falls on the unfilled casting before the last solidification. This happens in light of the fact that the rate of evaporation of polystyrene pattern is higher than the metal flow rate. Pattern vaporizes in front of the flowing stream of the metal as a result of the high warmth generated from the hot metal. Accordingly, the sand falls in the pit and thus creates a flawed casting. (R.L. Timings, 2009)

This issue has been overcome by consolidating the procedures of EPC and vacuum sealed molding process.

Vacuum Assisted Evaporative Pattern Casting Process (VAEPC)

The idea of hybrid casting process is acquainted with a point to accomplish better execution of the advanced casting procedures. The purpose behind creating hybrid-casting technique is to make utilization of joined or commonly improved focal points and to keep away from or diminish some antagonistic impacts of the constituent procedures when they are independently used. The Evaporative pattern casting process (EPC) is a without a binder process wherein no physical holding is required to tie the sand particles. The procedure in any case, does not ensure a sound casting each time as it is noticeably hard to control the rate of pouring in order to coordinate with the rate of polystyrene consuming. To defeat this impact, EPC is consolidated with the Vacuum Sealed Molding process. In the Vacuum Sealed Molding process, the physical holding is accomplished by utilizing pressure or vacuum in the molds. The vacuum framework when utilized with EPC helps in pulling back the decayed gasses and it additionally encourages in giving unbending nature to the sand shape. Additional (other than EPC process) requirements for Vacuum Assisted Evaporative Pattern Casting Process (VAEPC) For a VAEPC process we do not require any special pattern materials and mold boxes. Only a vacuum arrangement is required.

The Pattern Box

The pattern box is presented in Fig. The pattern plate is set on the straight top open surface of the pattern box and ideal setting between the mating surfaces is accomplished. Keeping in mind the end goal to find out the ideal setting, the pattern plate is furnished with four helping obstructs at the four corners in the pattern box. Extra support at the inside has been given to avert twisting of the pattern plate when subjected to vacuum pressure. The pattern box is settled to the vibrating table.pattern plate, having little gaps at equidistant spots can be utilized.

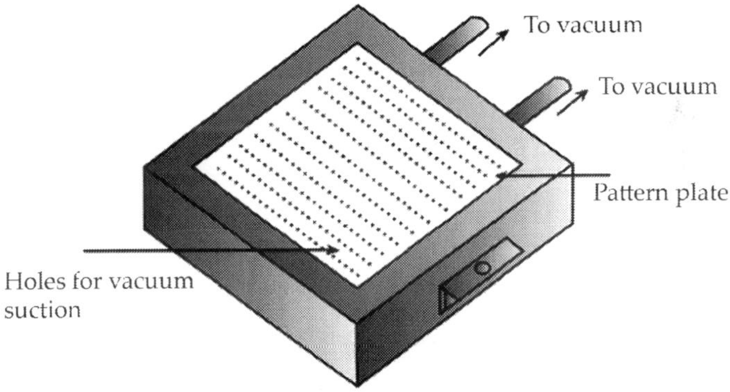

Figure 36 Pattern box for VAEPC process (nptel.ac.in, n.d.)

Molding Box

For Design of mold box, the real thought is to have a uniform circulation of vacuum all through the assemblage of sand to shape the mold. The plan of mold box for VAEPC process is more complex than traditional sand shaping procedure. The box must be made with annular divider on all the four sides. In side dividers are given windows. These windows are secured with fine mesh and upheld by a metal strip having little openings for supporting the mesh. This mesh keeps fine sand particles from being sucked up from the sand voids in the box and into the vacuum pump. Two pipes are associated with outside mass of the box. One pipe sucks the vacuum from the box and another pipe discharges the vacuum to the air. The mold box is presented in Fig

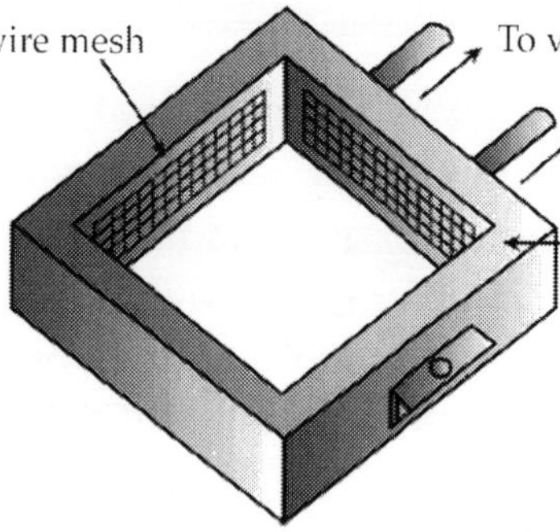

VAEPC: Process Description

The procedure steps of VAEPC are presented in Fig.

- The VAEPC method starts with the pre-development of beads, normally polystyrene. After the pre-extended beads / globules are balanced out, they are blown into a shape to frame pattern segments. In the shape, the steam cycle helps in combining the globules together by full extension.
- The distinct segments of pattern are consolidated by sticking and collected. The gating framework too is stuck and appended in a comparable manner.
- A ceramic coating covers the foam pattern cluster and it frames an obstruction with the end goal that entrance of liquid metal or disintegration of sand by pouring does not occur.
- Once the coating is dried, the assembly is set into a box containing bonded sand as the supporting material. In this procedure, any sand which has enough refractory quality to oppose the liquid metal temperature can be utilized. Silica sand, zirconia sand, olivine sand, and chromite sands can be utilized as mold sands. As in the EPC procedure, the sand loss is less and quite a bit of it is recuperated, even the costly sands, for example, zircon or chromites can be utilized. The frictional resistance between the sand grains decides the quality of the shape. The mold quality with precise grains is higher, albeit higher mass thickness is given by the adjusted grains. (This progression is the same the initial step of EPC)
- The pattern, covered with a suitable refractory wash, is

additionally inserted with dry, unbounded sand. It is next vibrated to create an inflexible shape.

- Two plastic films are utilized to embody the shape in the middle of and vacuum is connected in the sand box. The vacuum offers unbending nature to the form and more noteworthy hardness is subsequently accomplished. After this procedure, the bunch is stuffed and vacuum is proceeded with along these lines making the shape prepared for pouring.
- After pouring the liquid metal sets correctly with the shape and size of the pattern and takes up the majority of its geometry. Thus the surface finish and dimensional precision of the pattern is imperative as it has an immediate connection to the casting quality being delivered.
- Once the casting is expelled from the box by shake out technique, almost no fettling is required, as the procedure doesn't require shape joint lines and even the cores are totally wiped out.

Process Parameters of VAEPC

To recognize the critical process factors influencing the nature of Vacuum Assisted Evaporative Pattern Casting (VAEPC) process, a Cause and Effect diagram is constructed and is presented in Fig

The process factors of the VAEPC process can be put in the accompanying classes:

- Factors in light of molding sand: Sand type, state of the sand grains, grain size and size spread.
- Factors in light of pattern: Density and size of pattern globules.
- Factors in light of coating of patterns: material utilized for the slurry and coating thickness.

Factors in light of vibration: Amplitude of vibration, frequency and time. Factors in light of vacuum: Degree of vacuum forced.

CHAPTER 4

Material Processing Techniques

INTRODUCTION

Materials processing is characterized as the arrangement of steps or "unit operations" utilized as a part of the fabrication of crude materials into completed merchandise products. The operations include a progression of modern processes with different mechanical or compound methodologies, for the most part used to deliver in huge amounts or clusters of products.

Many other products may come about contingent upon the processes used amid the fabrication of a completed product. Numerous modern processes create dangerous by-products that may require unique handling while others can really be utilized to produce extra profits.

The processes utilized for the making of a coveted product for the most part serve two noteworthy tasks: the forming or changing of materials. The development of parts or products is normally accomplished by molding, casting or the use of stress. The physical state or form of items and products can be adjusted by various diverse means including chemical, mechanical and heating operations.

SHAPE TUBE ELECTROLYTIC MACHINING (STEM)

Shaped Tube Electrolytic Machining (STEM) is founded on the suspension process when an electric potential difference is enforced between the anodic work piece and a cathodic tool. An electrolyte (Often sulfuric Acid) causes the anodic work piece surface to detach in the presence of electric field. Afterward the detached metal ions are dissolved in the solution, they are removed by electrolyte flow as shown in Fig. . In this process the tool is a conducting cylinder having an insulating on the outer side and moved toward the work piece with a certain feed rate of electrolyte at the same time a voltage is applied across the machining space (McGeough, 1998). In this way the desired shape is obtained.

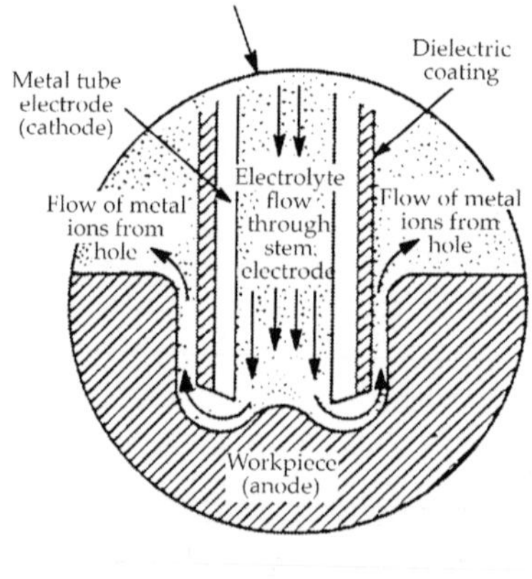

Figure 4.1 Illustration of Shape Tube Electrolytic Machining (STEM) process (Anon., n.d.)

STEM is, therefore, an improved deviation of the ECM that practices acid electrolytes. According to Rumyantsev and Davydov (1984), the process is proficient to produce small holes with diameters of 0.76 to 1.62 mm and a depth-to-diameter ratio of 180:1 in electrically conductive materials. With ordinary ECM Machining process it is very hard to machine such small holes because the insoluble precipitates causes obstruct to flow path of the electrolyte. (Black, 2015)

The machining system arrangement is alike to that used in ECM. However, it must be of less rigidity, be acid resistant, and have a periodically reverse polarity power supply. The cathodic tool electrode is made of titanium, its outer wall having an insulating coating to permit only forward machining of the anodic work piece. The usual working voltage is 8 to 14 V dc, while the machining current touches 600 A. in Metals Handbook (1989) it is reported that when a nitric acid electrolyte solution (15% v/v, temperature of about 20°C) is pumped through the gap (at 1 L/min, 10 V, tool feed rate of 2.2 mm/min) to machine a 0.58-mm- diameter hole with 133 mm depth, the resulting diametric overcut is 0.265 mm, and the hole conicity is 0.01/133.

10% concentration sulfuric acid is used to prevent the sludge from clogging the tiny cathode and guarantee an even flow of electrolyte through the tube. A periodic reversal of polarity, typically at 3 to 9 s prevents the gathering of the undissolved machining products on the cathode drill surface. The reverse voltage can be in the range of 0.1 to 1 times the forward machining voltage. In contrast to the EDM, EBM, and LBM processes, no heat-affected layer is produced in STEM, which is responsible to develop micro cracks. (Bralla, 2007)

Process Parameters

Electrolyte

Type	Sulfuric, nitric, and hydrochloric acids
Concentration	10 25% weight in water
Temperature	38°C (sulfuric acid) 21°C (others)
Pressure	275-500 kPa
Voltage	
Forward	8 14 V
Reverse	0.1 1 times the forward
Time	
Forward	5 7 s
Reverse	25 77 ms
Feed rate	0.753 mm/min

Process Capabilities

Hole size 0.5 6 mm diameter at an aspect ratio of 150

Hole tolerances 0.5-mm diameter ±0.050 mm

1.5-mm diameter ±0.075 mm

60-mm diameter ±0.100 mm

Hole depth ±0.050 mm

Acid electrolytes is used in this process that's why its use is narrow to drilling holes in only stainless steel or other corrosion-resistant materials in jet engines and gas turbine parts such as,

- Turbine blade cooling holes
- Fuel nozzles
- Any holes where EDM recast is not desirable
- Starting holes for wire EDM
- Drilling holes for corrosion-resistant metals of low conventional machinability
- Drilling oil passages in bearings where EDM causes cracks.

Shape of turbulator that are produces by intermittent drill have improved geometry due to STEM. Turbulators are mostly employed for enhancing the heat transfer in turbine engine-cooling holes.

Advantages

- The depth and diameter ratio of about 300.
- A huge amount of holes of about 200 in number can be drilled in a single pass.
- We can machine many nonparallel holes.
- We can easily drill the Blind holes.
- There are no resulting recast layers and material flaws. Complex shaped holes and many curves and slots are easily made by this process.

Limitations

- The process is used for corrosion-resistant metals. STEM is not fast if we need to drill only single holes.
- A special workplace and environment are required when handling acid. toxic waste is produced.
- Complex machining and tooling systems are required.

ELECTROLYTIC IN-PROCESS DRESSING:

A traditional method used for dressing a metal bonded grinding wheel during an accurate grinding process is Electrolytic In-Process Dressing. It is a new technique used for dressing harder metal-bonded super abrasive grinding wheel during grinding. ELID eradicates the wheel loading issues, that's why this is a hybrid process. ELID is the mixture of an electrolyte process and a mechanical process; hence any change in any one process may have a strong influence on the other. The uncertainties happened during the selection of an electrolytic parameters for dressing, the lack of knowledge of electrolytic mechanism of the ELID-grinding wheel etc., are decreasing the wide spread use of ELID process in the manufacturing industries. (Black, 2015)

Principle of ELID

In Electrolysis process electrical energy is converted into chemical energy. The process occurs in an electrolyte, which ions are transfer between two electrodes. The electrolyte is the conducting media between the two electrodes which are also connected to a direct current as shown in Fig. 3, and the unit is known as electrolyze cell. When electricity is supplied, the positive ions move towards the cathode at same time the negative ions will move towards the anode. Positive ions are called cations and are normally metals. Because metals have tendency to lost electrons and are able to pick up electrons. Anions are negative ions. They carry more electrons than normal and have the chance to give them up. If the cations have interaction with the cathode, they get the electrons they lost back to become the elemental state. The anions react in a vice versa when they in contact with the anode. They hand over their redundant electrons and become the elemental state. Therefore the cations are reduced and the anions are oxidized. To control the reactions in the electrolyze cell various electrolytes (the electrolyte contains the ions, which conduct the current) can be chosen in order to stimulate special reactions and effects. The ELID work on similar principle but the cell is varied by using different anode and cathode materials, electrolyte and the power sources suitable for machining conditions. (K. G. Swift, 2013)

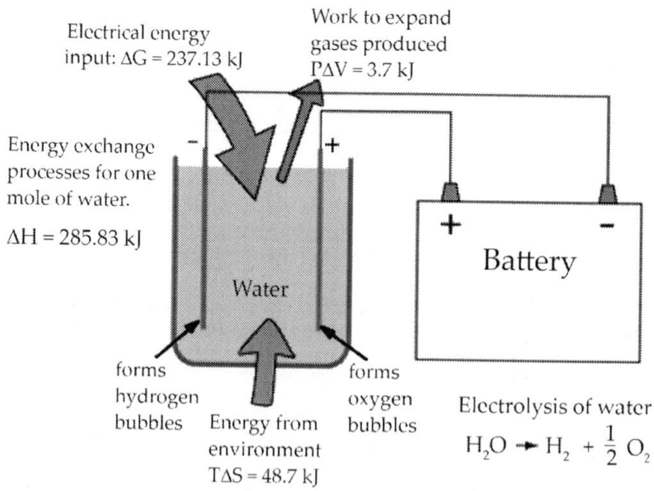

Figure 4.2 Illustration of an electrochemical cell (Anon., n.d.)

The cell consists of a conductive wheel, an electrode, an electrolyte and a power supply, which is known as the ELID system. Figure shows the schematic illustration of the ELID system. Grinding wheel is metal-bonded into a positive pole with the help of brush smoothly contacting the wheel shaft. The electrode is made into a negative pole. In the small gap of approximately 0.1 to 0.3 mm between the positive and negative poles, electrolysis occurs through the supply of the grinding fluid and an electrical current.

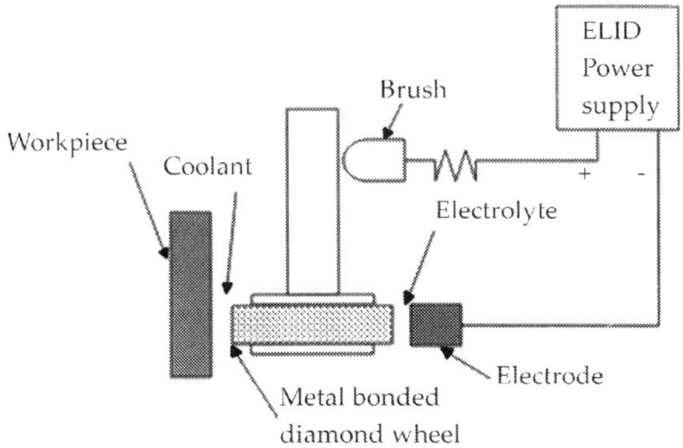

Figure 4.3 Schematic illustration of the ELID system. (Kumar, n.d.)

Conducting materials are used to make ELID grinding wheel i.e. metals such as cast iron, copper & bronze. The diamond layer is set by mixing the metal and the diamond grits with certain volume percentage, and the wheels were prepared by powder metallurgy. The prepared diamond layer is attached with the steel hub. The grinding wheels are available in different size and shapes. Among them the straight type and the cup shape wheels are usually used.

The function of the Electrolyte

In-Process dressing process the most important thing is electrolyte. The efficiency of ELID depends upon the properties of electrolyte. There will not be any oxide layer on the wheel surface if the oxide layer produced due to electrolysis is solvable. Faraday's Law is responsible for material oxidized from the wheel surface. However in ELID, electrolyte is used that doesn't dissolve oxide and therefore metal oxides are deposited on the grinding wheel surface during in-process dressing. Ohmori et al studied the performance of different electrolyte in ELID, which shows the importance of selection of the electrolyte. Water is used to dilute (2%) the electrolyte and used as coolant and electrolyte for grinding. Some amount of chlorine is added to water because it has a positive potential, which has a significant effect on electrolysis.

Power Sources

AC, DC & Pulsed DC have been used practically in ELID. Ohmori's (1995, 1997) uses different power sources to compare the uses and the advantages experimentally and showed results. According to modern research it is clear that pulsated power sources can provide better control on the dressing current than other power sources. When ELID use DC-pulsed power source, it is important to know the fundamentals of pulsed electrolysis to achieve enhanced performance and control

Different Methods of ELID

ELID consist of four major groups based on the materials to be ground and application of grinding, even though the principal of in-process dressing is analogous for all the methods. The different methods are as follows:

- Electrolytic In-process Dressing (ELID I),
- Electrolytic Interval Dressing (ELID II),
- Electrolytic Electrode-less dressing (ELID III)
- Electrolytic Electrode-less dressing using alternate current (ELID IIIA)

Separate electrode is used according to conventional and most common studies. The basic components of ELID are power supply, a metal bonded grinding wheel and an electrode. The electrode used could be ¼ or 1/6 of the perimeter of the grinding wheel. Usually copper or graphite could be best for electrode materials. 0.1-0.3 mm gap is adjusted between electrode and grinding wheel. Efficiency of process depends on the suitable gap and coolant flow rate. Generally curve fashioned electrode are used in this type of Process and the wheel used is either conventional type.

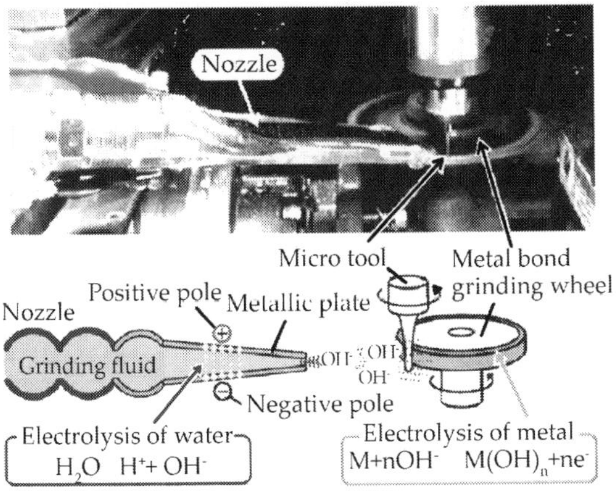

Figure 4.4 ELID Grinding and EDM for Finish Machining (MIMechE, n.d.)

Electrolytic Interval Dressing (ELID II)

In most of the industrial fields machining of small hole in hard and brittle materials is highly required. The problem in micro-hole machining includes the following:

- Difficult to prepare small grinding wheels with high quality,
- Calculation of grinding wheel wear compensation and
- Accuracy and surface finish of the holes are not satisfactory.

Due to difficulty of mounting of an electrode, the current ELID grinding process is not appropriate for micro-hole machining. Using the combination of sintered metal bonded grinding wheels of small diameter, Electric Discharge Truing (EDT) and Electrolytic Interval Dressing (ELID II) could solve the problems in micro-hole machining. The smallest grinding wheel for example 0.1 mm can also be trued accurately by using EDT method, which uses DC-RC electric power. With the help of electrolysis, tiny grinding wheels can be pre-dressing in order to achieve better grain lumps. The dressing parameters should be selected carefully to avoid unnecessary wear of grinding wheel. The grinding wheel is dressed at a fixed interval based on the grinding force. If the grinding force increases beyond certain threshold value, the wheel is re-dressed. (Golam Kibria, 2017)

Electrode-less In-process dressing (ELID III)

Wheel loading and clogging is increasing due to the entrenching of turnings on the grinding wheel surface and reduces the wheel efficiency for grinding of material such as steel. The efficiency of the grinding wheel is increased if the size of swarf is smaller. For machining of conductive materials like hardened steels, metal-resin-bonded grinding wheels have been used. The conductive work piece acts as the electrode and the electrolysis arises between the grinding wheel and the work piece. Usually the bonding material used for grinding wheel is copper or bronze. The electrolytic layer is made on the work piece and it is detached by the diamond grits. Thus the chips production is controlled by using electrode-less in-process dressing (ELID III). The base material is oxidized and the wheel surface contains resin and diamond grits, during electrolytic dressing. Theoretically the metal bond is removed by electrolysis, but practically the grinding wheel surface contains voids, which is caused due to electric discharge. When high electric parameters are chosen, the amount of electric discharge increases and it causes destruction on both the wheel and ground surfaces. For improved surface finish, low voltage, low current, low duty ratio and low in- feed rate should be selected.

Electrode-Less In-Process Dressing using Alternative Current (ELID IIIA)

ELID –IIIA resolve the difficulties of using electrode-less in-process dressing. The AC produces a thick oxide film on the surface of work piece, which avoids the direct contact between grinding wheel and the work piece. Hence the electric discharge between the wheel and work piece is completely eliminated and the ground surface finish is improved. (Golam Kibria, 2017)

The main purpose of ELID is to provide continuous grinding using harder metal-bonded wheels. The difficulties such as wheel loading and glazing can be eradicated by introducing an electrolyze cell (anode, cathode, power source and electrolyte) during grinding, which stimulates electrolysis whenever necessary. The electrolyze cell required for the in-process dressing is different from the cell used for standard electrolysis or electroplating. Therefore, attention should be focused on the selection of factors such as the bond-material for the grinding wheels, electrode material, the electrolyte and the power source. If any one of the parameters is not selected properly, the result obtained from the electrolysis will be different. Therefore, an sufficient knowledge about the electrolysis is necessary before incorporate with the machining process. This chapter provides the necessary information about the ELID, selection of bond material for the ELID, the electrode material selection for the grinding wheels, electrolyte and the power source selections.

Application

- The structural ceramic components
- Bearing steel
- Chemical vapor deposited silicon carbide (CVD- SiC) Precision internal grinding
- Mirror surface finish on optical mirrors
- Micro lens
- Form grinding
- Die materials
- Precision grinding of Ni-Cr-B-Si composite coating
- Micro-hole machining ELID-lap grinding of silicon wafers

ELECTROCHEMICAL GRINDING

In Electrochemical grinding (ECG) use a negative charged abrasive grinding wheel, electrolyte solution, and a positively charged work-piece, as shown in Fig.7. This process is similar to ECM only deviate in cathode that is a specially constructed grinding wheel instead of a cathodic shaped tool like the contour to be machine by ECM. Diamond and Aluminum oxide are the insulating abrasive material on grinding wheel is set in the conductive bond material. But ECG uses the non-conducting abrasive particles which act as spacer between the wheel conductive bond and the anodic work piece. A constant inter-electrode clearance (less than .025mm) through which the electrolyte is flushed can be kept constant, this depends upon the grain size of the particles. (Black, 2015)

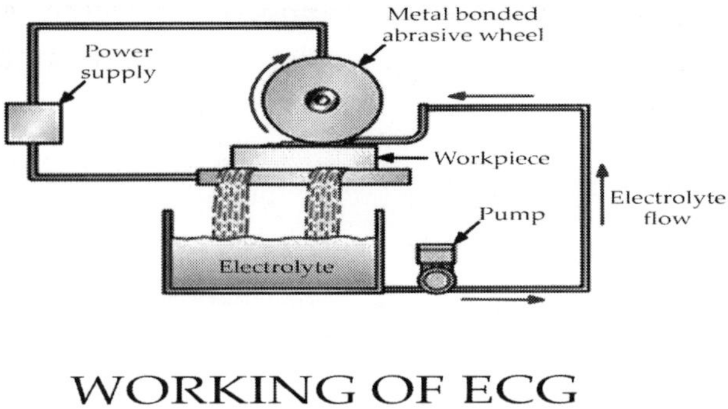

Figure 4.5 Illustration of electro chemical grinding (Anon., n.d.)

Machining products are to be removed continuously from the working area by abrasives. Cathodic tool is the rotating wheel with abrasive particles (60 320 grit number) on its circumference as shown in Fig. . Electrolyte flow, normally NaNO3, is provided for ECD. In ECG the grinding wheel rotate at the speed of 20 to 30 ms-1 with current is in the range of 50-300 A.

Material Removal Rate

When potential difference of 4 to 40 V is applied between the grinding wheel (Cathodic) and the work piece (Anodic) having current density of about 120 to 240 Acm-2. Material to be machined, clearance width and applied voltage are dependable to the current density. ECD mainly remove the material, while the MA of the abrasive grits removes for an further 5 to 10 percent of the total material.

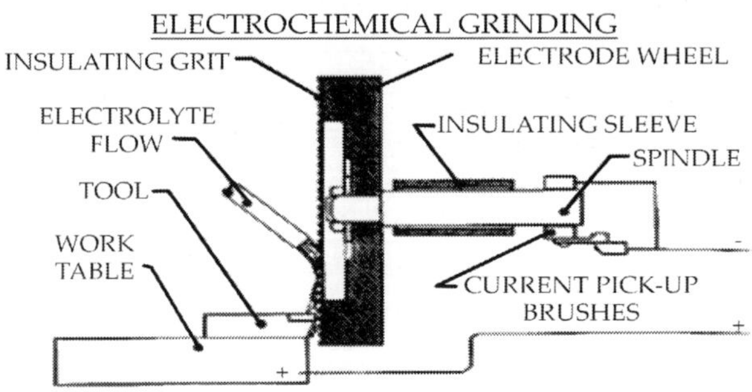

Figure 4.6 illustration of electro chemical grinding and machining system (Anon., n.d.)

ECG produces burr free parts with no stressed and material removal rate is 4 times faster than conventional grinding. Removal rate is measured in volumetric removal rate (VRR), in case of ECG VRR is 1600 m3/min. According to McGeough (1988) and Brown (1998), maximum removal rate can be obtained when grinding area should be as large as possible to pull greater machining current, which affects the ECD phase. Volumetric removal rate can be calculated by following equation.

$$\text{VRR} = \frac{\varepsilon I}{\rho F}$$

Where,

ε = equivalent weight, g

I = machining current, A

ρ = density of work piece material, g/mm3

F = Faraday s constant, C

ECG consists of both MA and ECD, so it is a hybrid machining process. This process have many advantages like higher machining rate, improved surface layer properties and reduced tool wear and energy consumption. ECD phase follows the Faraday's laws. Conditions in gap such as electric field, flow of electrolyte and hydrodynamic effects on boundary layers near anode are responsible for the action of abrasive grains. Figure 9 shows the basic components of the ECG process. Material removal & surface layer formation by these two machining process depends upon the process parameters. The involvement of each machining phase to the material removal from the work piece has resulted in a significant increase in the total removal rate QECG, in relation to the sum of the removal rate of the electrochemical process and the grinding processes QECD and QMA, when keeping the same values of respective parameters as during the ECG process. (Benedict, 1987)

The grinding process of MA is illustrated in Fig. 10 with a rotary conductive abrasive wheel, improved ECD process. Abrasive action of grains can be performed by mechanical depolarization by abrading the possible insoluble films from the anodic work piece surface. Alloys of different metals and cemented carbides results such types of films. A special purpose of abrasive grains is to depassivate mechanically the surface of work piece. In the machining region there is an area of common ECD and MA of the work piece surface, where the gap width is less than the height of the grain part projecting over the binder. A area of totally electrochemical removal where the abrasive grains do not in contact with the work piece surface occurs at the entry and exit sides of the wheel.

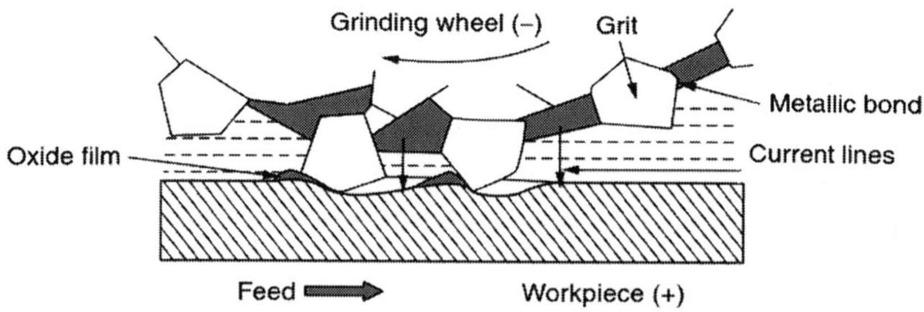

Figure 4.7 ECD and MA in the machining gap during ECG. (Anon., n.d.)

Process Characteristics

- In ECG process the grinding wheel life is very long as up to 90% of the metal is detached by electrolysis action. Only abrasive action diminishes life of grinding wheel.
- ECG process efficiently produce very smooth and burr free edges as compared to conventional mechanical grinding process.
- Work piece bending is less because much less heat is produced during ECG process.
- Very little tool and work piece contact because majority material removal in ECG process is transferred by dissolving action through chemical process. This make ECG ideal for following parts machining.
- Brittle work-pieces which otherwise are very hard to grind by the conventional process
- The parts that cannot resist thermal damages and
- The parts designed for strain and burr free applications.

Applications

- The ECG process can be applied effectively
- Machining parts manufactured from hard-to-cut materials like sintered carbides, creep-resisting (Inconel, Nimonic) alloys, titanium alloys, and metallic composites.
- Applications alike to milling, grinding, cutting off, sawing, and tool and cutter sharpening.
- Manufacturing of tungsten carbide cutting tools, fragile parts, and thin- walled tubes.
- Elimination of fatigue flaws from steel structures under seawater. In such an application holes about 25 mm in diameter, in steel 12 to25 mm heavy, have been produced by ECG at the ends of fatigue cracks to stop further expansion of the cracks and to allow the removal of specimens for metallurgical inspection.
- Producing specimens for metal fatigue and tensile tests.
- Machining of carbides and a variety of high-strength alloys.
- The ECG process can be useful for general methods of grinding
- Face wheel grinding,
- Cone wheel grinding,
- Peripheral or surface grinding,
- Form wheel or square grinding.
- The process is not modified to cavity reducing, and therefore it is inappropriate for the die- making industry.

Advantages

- Lack of work hardening
- Removal of grinding burrs
- Lack of bend of thin brittle or thermo sensitive parts
- Good surface quality
- Production of thin tolerances
- Longer grinding wheel life

Disadvantages

- Costly need huge investment
- Process narrow to electrically conductive materials
- Corrosive nature of electrolyte
- Requires disposal and filtering of electrolyte

ELECTRO-CHEMICAL ETCHING (ECE)

Etching is used to remove material. Work piece is dipped in an etchant which removed material by chemical action on the surface area in contact with etchant. Normally in this process the metal is converted to its salt that dissolved in the etchant and hence removed from surface. Material is withdrawn from the etchant after removal of desire amount of material and the process is stopped by washing. Etching is frequently done selectively, by coating surface areas that are to be secure and leaving other are as exposed for etching. The coating may be previous applied film of material i.e. SiO2 or etch resistant.

In semiconductor manufacturing this process is classified in two main groups:

- Wet chemical etching (Older but easier)
- Dry plasma etching (have certain limitations)

Wet Chemical Etching

An aqueous solution is used in wet chemical etching normally an acid in water attack the work piece. Etching solution is selected according to work piece because an etchant can attack on a certain material and no protective film form during process. It is very simple process, masked wafers are immersed in the suitable etchant for specific time and thereafter quickly washing the work piece to stop the etching process. The amount of material removed depends on the process parameters such as immersion time, etchant concentration & temperature. (Company, 2004)

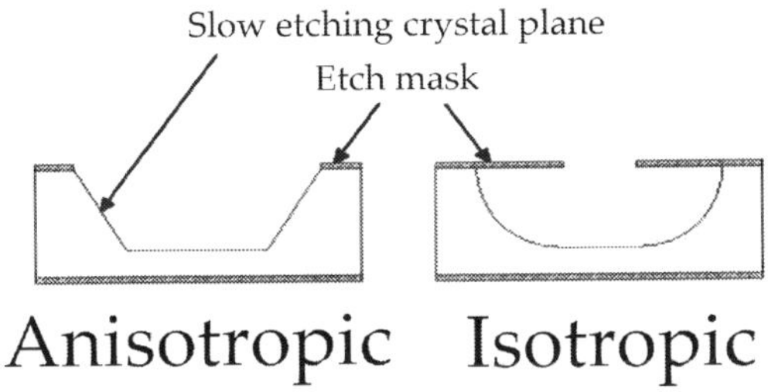

Figure 48 Illustration of etching process (Anon., n.d.)

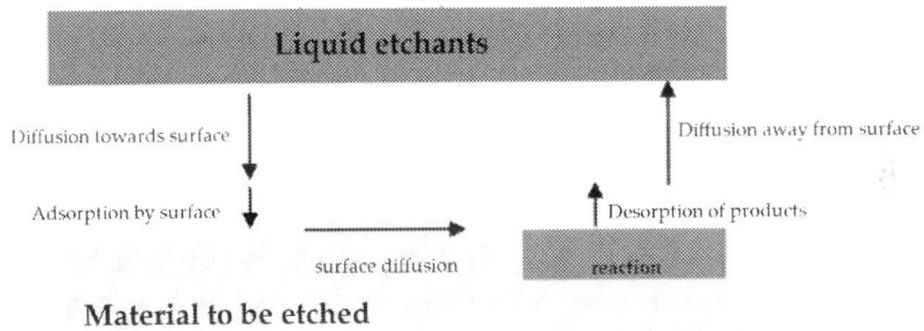

Figure 49 Wet Etching Mechanism (Anon., n.d.)

Fig. illustrate the etching process there are two types of reaction anisotropic and isotropic. As shown in Fig. isotropic reaction occur in all direction resulting a cavity under the protective mask while in anisotropic etching the reaction occur in only one direction. Wet chemical etching is widely used and mask pattern must be according to size to control the etching.

Keep in mind that the etchant does not react with layer under the desire material to be removed. Etchant is prepared having such composition that will only react with the desire metal, not with other material in contact with it. In general practice other materials are also exposed to etchant, but the attack is less than the targeted material. The ratio of etching rates in objective metal and other materials define the etch selectivity, i.e. etch selectivity of HF is infinite for SiO2 as compared to Si. (Jr, 2011)

The etch selectivity of the etchant is the ratio of etching rates between the target material and some other material, such as the mask or substrate material. For example, etch selectivity of hydrofluoric acid for SiO2 over Si is infinite. If process control is insufficient, either under-etching or over-etching can occur, as in Figure 12. In under etching objective layer is not entirely removed, results when the etching time is too small and/ or the etching solution is too weak. Over-etching consists excessively removal of the objective material, subsequent in loss of pattern definition and possible damage to the layer below the objective layer. Over-etching is began by overexposure to the etchant.

Advantages

- Cheap
- Consistency
- High Throughput
- Excellent Selectivity

Disadvantage

- Very hard to control Critical feature Dimension
- Difficult to control the degree of over etching due to undercut
- Decrease in Etch rate as Reagent solutions are consumed
- Hazardous and Difficult to handle
- Toxic Fume

Applications of Wet Process

Silicon Oxide Etch

$SiO_2 + 6HF \rightarrow H_2SiF_6 + 2H_2O$

HF: Etchant, NH4F: Buffering Agent

Poly-Si Etch

$Si + HNO_3 + 6HF \rightarrow H_2SiF_6 + HNO_2 + H_2 + H_2O$

HNO_3: Oxidant, HF: Etchant, CH_3COOH: Buffering Agent

Al Etch

HNO_3: Oxidant, H_3PO_4: Etchant

Silicon Nitride Etch

Hot (>150°C) H_3PO_4: Etchant

DRY PLASMA ETCHING

Ionized gas is used to etch the target material in dry plasma etching. To create plasma mixture of suitable gases are introduced in vacuum chamber and electrical energy to ionize the mixture.

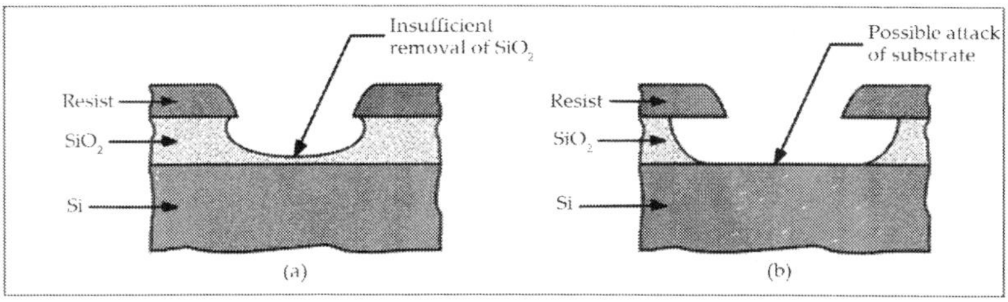

Figure 4.10 Two problems in etching: (a) under-etching and (b) over-etching (Anon., n.d.)

The high-energy plasma chemically attacks the objective surface, evaporating the material to eradicate it. There are numerous ways in which plasma can be used to etch a material; the two principal processes in IC fabrication are plasma etching and reactive ion etching.

Ionized gas produces atoms or molecules that are chemically reactive; these reactive atoms etch the target material chemically on exposure. The plasma etchants are generally created with fluorine or chlorine gases. In plasma etching etch selectivity problem is more than wet chemical etching i.e. SiO2 shows best etch selectivity over Si in typical plasma etching process as compared with limitlessness with HF Chemical Etching.

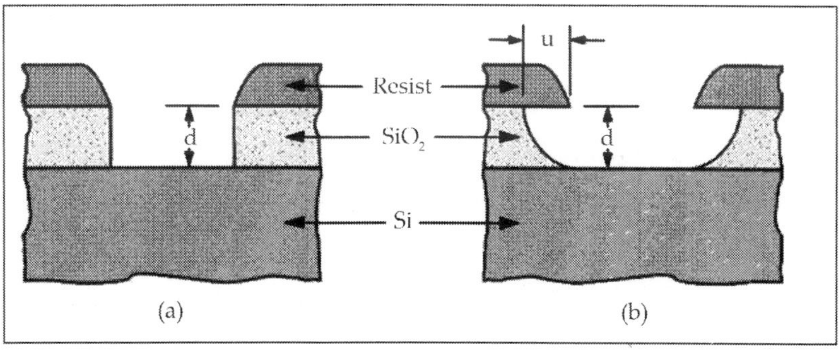

Figure 4.11 (a) A fully anisotropic etch, with A=1; and (b) a partially anisotropic etch, with A=approximately 1.3. (Anon., n.d.)

Another purpose of the ionized gas can be to materially attack the target material, producing atoms to be emitted from the surface. This is the process of sputtering, one of the techniques in physical vapor deposition. When used for etching, the process is called sputter etching. This procedure of etching has been applied in semiconductor processing and it is more common to unite sputtering with plasma etching as demonstrated above, this causes a phenomenon known as reactive ion etching. As a result of this process both chemical and physical etching produces on the work piece.

The benefit of the plasma etching processes against wet chemical etching is that they are much more anisotropic. This property can be freely defined with reference to Figure 13. In (a), a fully anisotropic etch is shown; the undercut is zero. The degree to which an etching process is anisotropic is defined as the ratio:

$A=d / u$

Where

A=degree of anisotropy;

d=depth of etch, (thickness of the etched layer)

u = the undercut dimension, as illustrated in Figure 4.3(b).

In wet chemical etching usual yield of is approximately 1.0 which means isotropic etching.

Plasma etching and reactive ion etching have large of anisotropy, but under those attained in sputter etching. As IC article sizes continue to contract, anisotropy becomes increasingly significant for succeeding the required dimensional tolerances.

CHEMICAL ETCHING

Chemical etching used free radicals in etching

Free radicals have no charge but have partial bonding (unpaired electrons)

i.e.

$e\text{-}1 + CF_4 \rightarrow CF_3 + F\text{-} + e\text{-}1$

Both F- and CF3 are free radicals.

Chemical reactivity is high

F wants 8 electrons rather than 7 and reacts rapidly to find a mutual electron.

The idea is to catch the free radical to react with the material to be etched (Si, SiO_2).

The derivatives should be gaseous so that it can be transported away.

The reaction below is such a reaction,

$4F + Si \rightarrow SiF_4$

There are frequently numerous more complex transitional states.

Addition of Gas additives can increase the production of the reactive species (O2 in CF4)

The chemical component of plasma etching occurs isotropically.

This is because,

The influx angles of the species is isotropic

There is a low sticking coefficient (which is more important)

The arrival angle follows what we did in deposition and there is a cosnq dependence where n=1 is isotropic

The sticking coefficient is,

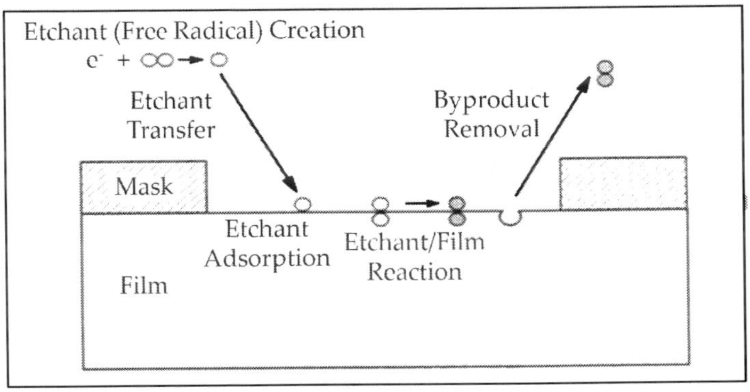

Figure 4.12 Process involved in chemical etching during plasma etch process (Anon., n.d.)

- A high sticking coefficient means that the reaction takes place the first time the ion raids the surface.
- For lower sticking coefficients, the ion can vacation the surface

(usually at random angles) and attacks the surface wherever else.

- One would deduction that the sticking coefficient for reactive ions is high.
- However, there are often complex reactions chained together. This difficulty often means low sticking coefficients Sc for O2/CF4 on Si is about 0.01
- This additional "bouncing around" of the ions leads to isotropic etching.
- Meanwhile free radicals etch by chemical reaction with the material to be scratched, the process can be highly selective

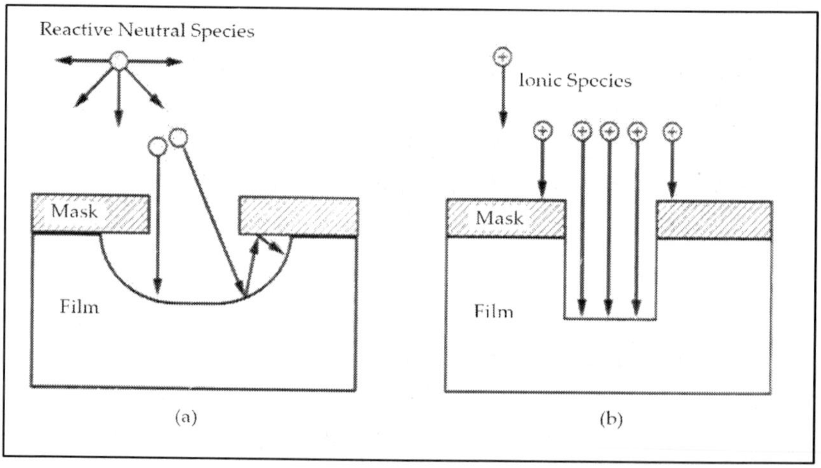

Figure 4.13 Fluxes of species in plasma etching: (a) fluxes of reactive neutral chemical species (such as free radicals), with a wide arrival angle distribution and sticking coefficient; (b) fluxes of ionic species, with a narrow, vertical arrival angle distribution (Anon., n.d.)

Advantages

- Isotropic
- Purely Chemical Reaction
- High Pressure
- Batch Wafer Type

- Less Electrical Damage
- Doe require as high doping level as boron
- Better thickness control
- Less stress on wafer
- Batch fabrication

ELECTRO JET MACHINING

Electro Jet Drilling

In modern technologies where size of everything is minimizing, this insisted us to use nontraditional machining techniques for small machining. For micro hole drilling a non-traditional method known EJD is used. This method is used in machining of Cooling holes in jet turbine blades, inkjet printers, printed circuit boards etc.

Working

In EJD a negatively charged flow of acid electrolyte is bombarded on the work piece to drill a hole. Concentration of acid electrolyte is 10-25 %, which is handled through a finely drawn tube nozzle under pressure (0.3-1.0 M/mm2). The electrolyte jet is charged with a platinum wire, inserted into the glass tube is fused to the negative terminal of DC power supply. The work piece acts as anode. When an appropriate electric potential difference is applied between the two electrodes, the material elimination takes place through electrolytic dissolution as the charged electrolyte stream attacks the work piece. When electrolyte flowed away from the surface of the work piece it removes the metal ions. A much stretched and diluent electrolyte flow path requires much higher voltage (150-750V) so as to affect adequate current flow. (Jr, 2011)

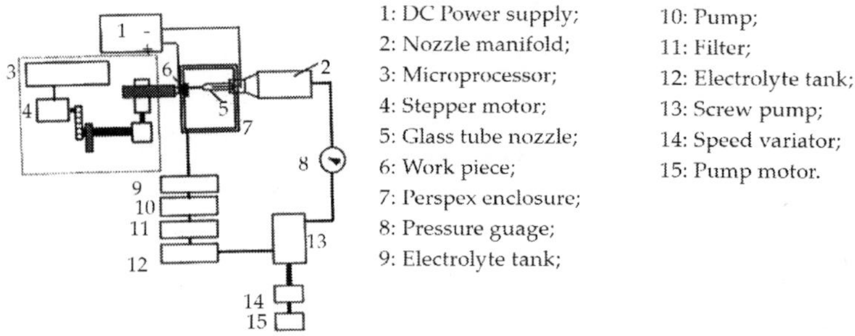

Figure 4.14 Schematic of experimental setup for electro jet drilling (Anon., n.d.)

Holes were machined by setting input parameters for each cycle. When the jet exits through work piece it is indication of ending of hole. An electronic timer is used to record the time spent during hole drilling. Work piece is weighed before and after drilling by electronic balance. Equation 1 is used to calculate the machining rate. Toolmarker microscope is used to measure the size of the hole. A sum of three diameter measurements were made at hole directions 60 degree separately and averaged values were used in calculations. Equation 2 is used to determine the overcut. Founded on the entry side hole diameter and outer side hole diameter measurements, the hole taper was calculated by equation 3.

$$MRR = \frac{Initial\ weight - Final\ weight}{machining\ time} \qquad (1)$$

$$overcut = \frac{d_{entry} - dglasstube - nozzle}{2} \qquad (2)$$

$$taper(\theta) = \tan^{-1}\left[\left(\frac{d_{entry} - d_{exit}}{2t}\right)\right] \qquad (3)$$

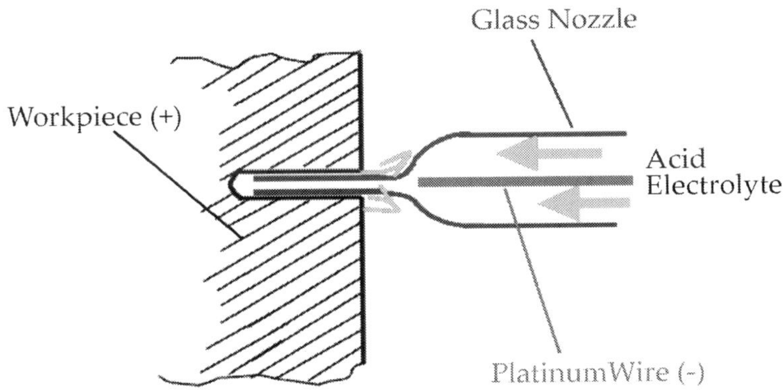

Figure 4.15 Working Principle of electro jet machining (Anon., n.d.)

MRR

Applied Voltage – according to Faraday's law MRR is increased due to increase in voltage and current. MRR rapidly increases above 350 Volts.

Electrolyte Concentration – MRR increases due to increase in electrolyte concentration because conductivity increases as a result of more current flow

Feed rate Increase in FR-reduces inters electrode slit- leads to smaller ohmic resistance-increases electrolyzing current.

Overcut

Applied Voltage – large overcut due to increase in voltage

Electrolyte Conc. Higher the concentration greater will be the overcut

Feed Rate less contact time due to high feed rate the greater will be overcut

Current overcut is increases as result of increase in current

Hole Taper

Difference between hole entrance and hole diameter will result the hole tapper. Hole tapper is greater when applied voltage and electrolyte concentration is increased because electro jet remains in contact with the entry side of the work piece for a longer time, larger entrance diameter than the exit diameter.

Advantages

- Micro-holes can be drilled
- Can be used for hard and brittle materials
- Cheaper than traditional drilling

Disadvantages

- Set up can vibrate otherwise hole will get big or distort.
- Maintenance cost is large

LASER BASED HEAT TREATMENT

Basic principle of laser beam heating is that it uses highly intense beam of coherent light focused on a pint. The beam is generally followed along a fixed path on the work piece surface, resulting heating of the steel into the austenite region. Machined area is immediately quenched as laser beam is passed away. Laser stands for "light amplification by stimulated emission of radiation". The benefit of LB above Electron beam EB heating is that laser beams do not need a vacuum to achieve best results. Energy density levels in Electron beam (EB) and LB heating are lesser than in cutting or welding. (Davis, 1995)

LB technology has directed to the possibility of restricted variations to the microstructures of a range of materials. This type of changes can improve service properties in the surface layers of a section but remaining properties are not changed in bulk material. Numerous methods for these changes but all depends on the ability of laser beam to manipulate accurately and high power density. Main advantages of laser Beam application are:

- Chemical purity and cosmetic look
- Less heat input, meanwhile the source temperature is so high, conversion occurs so rapidly and the heat input to the part is very little. This decreases the bending and the heat- affected zone is very minor.
- No need of post marching

- Process is non-contacting
- Integration is easy

Laser Heat Treatment Principles

Laser uses the same principle of heating as in conventional methods. Laser involve the time scale but magnitude is shorter. Conventional methods are used to heat like furnace, flame, arc or induction coil, on the other hand laser beam concentrated or shaped into a desire pattern and scanned over the component. It heats exterior more quickly due to high energy density, decreasing the time for the transfer into the massive component. Laser heat treatment and surfacing techniques must comprehensive directly with a broad variety of relatively low cost orthodox processes and must therefore offer important benefits.

A ray of energy is produced in laser in the form of continuous or fluctuated. The power density is one parameter of laser which includes power of beam and the focus diameter. Second one is the rate of power density movement across the surface. Usually the rate is expressed as interaction time i.e. the time duration for which the laser beam is focused on a point on surface. Fig. represents the advance version of heat treatment processes.

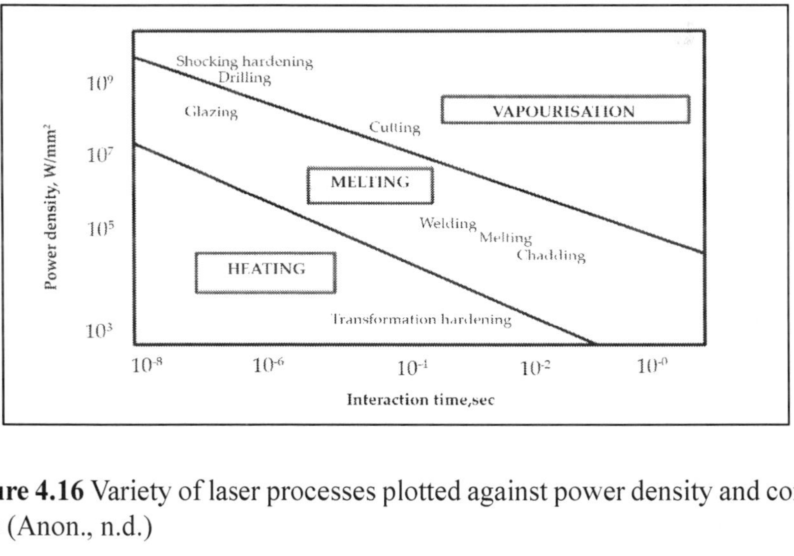

Figure 4.16 Variety of laser processes plotted against power density and contact time (Anon., n.d.)

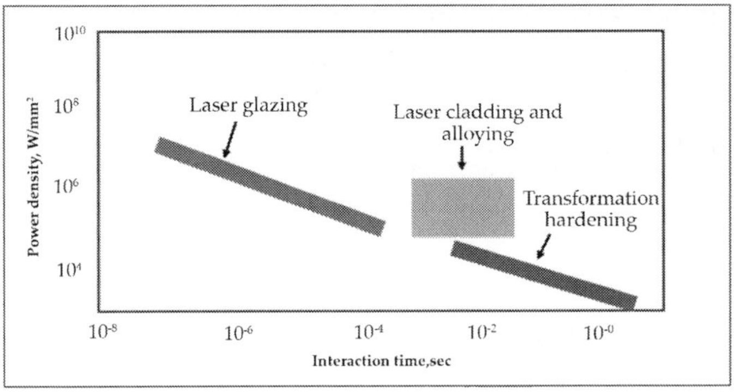

Figure 4.17 *Laser heat-treating rules (Anon., n.d.)*

Materials of high hardness may be handled with low power density and an enhanced interaction time, in order to complete a homogeneous case with substantial depth. Materials with low hardenability are treated with greater power density and minor interaction times in order to produce the quick cooling rates essential for martensite development at cost of a narrower case.

Laser Sources

In modern reign of technology there are many type of laser sources

- CO2
- Lamp and diode pump
- Nd: YAG and high power diode laser

These types are being used foe heat-treatment application. A decade ago only CO2 laser rays can produce a combination of power density and interaction time required for laser heat treatment. The invention of multikilowatt Nd: YAG lasers with flash lamp and diode pumping together make available a different source, with numerous advantages. One of the main benefit of the Nd: a YAG laser source is that the wavelength of the laser light (1.06 μm) permits the ray to be conveyed via an optical fiber with comparatively minor energy losses. This enable flexible distribution of laser rays at the processing head. So, Nd: YAG lasers producing high levels of laser power can be operated using robot, creating them perfect for three- dimensional processing.

For example the ray wavelength falls i.e. 1.06μm related to 10.6μm for CO_2 laser, the absorptivity of metal surfaces rises, and so an absorptive coating is no longer needed, hence simplifying the operation significantly. More freshly, multikilowatt diode lasers have been industrialized with wavelength of 0.8μm, which are dense and can be fixed directly on a robot for hardening of complex geometry components.

Beam Shape

A spot shape beam is repeatedly used for strengthening of CO_2 and Nd: YAG laser beams. This is produced by simply defocusing the rays, and is a suitable clarification for many engineering uses. The penetration profile of tough region can be approached to the mirror image of the beam power distribution, with lesser amplitude and some turning of the edges ensuing by lateral heat flow. By using rays shaping optics, the shape of the toughened sections can be diverse and may be conceivable to harden at higher exposure rates. If an even penetration profile of continuous width is needed, a kaleidoscope is the inexpensive solution.

Factors Affecting the Laser Process

Shield Gas

There are two main functions of shield gas in the laser heat treatments. It protections the heated/melt region from corrosion and also defends the concentrating optics from the vapors. Mixture of argon and nitrogen shield gases is usually used and usual flow rates are almost 20l/min. method of shielding and diameter of nozzle that is used to deliver the gas effect the flow rate.

Feed Rate

Power density and path thickness supplies fix the length of the beam travel direction. A power level in between 1-4kW is usually used. A high power allows high feed rate to be used, with respectively high exposure rates. However, the applied range that can be used significantly as danger of both hotness, leading to surface melting or a lacking peak temperature with no hardening. Feed rate is the flexible that is usually altered when fine-tuning the process in order to attain the compulsory hardened penetration and amount of homogenization.

System Basics

Laser processing system consists of three basic components; material handling, speed & controls, and laser light sources. These system elements are not distinctive when bearing in mind numerous materials processing cells used in industry today. Related with prime participants of laser technology, such as induction and flame hardening systems, the merely major variation is the use of a laser for the energy source.

Materials handling is a main attention in a laser-treating system like traditional material processing systems. The problem is more than just receiving parts in and out, as the economics of moving parts from one place to another can be an important factor. From physical systems used in job shops to complete mechanization on the engineering floor, the cost of materials handling can surpass that of indication and control and the laser collectively. When seeing a laser system, it is significant to determine as closely as likely what the laser on-time is for a specific product. The laser should be applied to treat parts at a duty cycle of 75% or better. Because distinctive cycles range between less than a second to providing 30 minutes, each application offers distinctive challenges.

Motion and control for laser systems frequently is extra refined than that mandatory for challenging technologies. Lasers are preferably suitable for computer control, proficient to turned on and off in a difficulty of milliseconds. It is not rare for a laser process to be organized to one-tenth of a second. Such accurateness needs close control of both of time and position.

Lots of cutting and welding systems are precise to 0.001 in. (0.02 mm) and repeatable to 0.0005 in. (0.01 mm). Although most heat treating uses do not need such close control, the potential happens to treat parts minor than the head of a pin with precision and repeatability. Moreover provided that unique control, laser hardening systems provide elasticity in that switching to another product often can be accomplished simply by selecting a new program and substituting tooling. Choice of the type of laser, the third basic component of a system, can be challenging for those not aware with the technology. The many existing choices boil down to three basic technologies: carbon dioxide (CO_2) lasers, neodymium: yttrium-aluminum-garnet (Nd:YAG) lasers, and high-power direct diode (HPDD) lasers. To be predictable choice was CO2 up to as newly as five years ago. However, a broader choice due the commercial readiness of both high-power Nd:YAG and HPDD lasers than the past a number of years has complex the selection process. (Black, 2015)

Lasers are useful for heat treatment having wavelengths between 800 and 10,600 nm. Above this variety of wavelengths, iron has almost a fourfold upturn in absorption. Since such poor absorption at extensive wavelengths, it is essential to alter the surface condition of a part to proficiently absorb light. This can be done by coarsening the surface, but the most general method is to put on an absorptive coating such as paints, inks, phosphates, oxides and oxyacetylene soot. Work piece temperature also disturbs absorption in service of laser heat treating. Outcomes show that using a wavelength of 1,060 nm, absorption by steel during laser heat treatment is about 60% and can be enhanced to 85% by the accumulation of a appropriate coating. At 10,600 nm, absorption can beat 70% when coatings are useful to a surface. Coating application can be mechanized and performed in the laser cell with slight to no drying time, as in the case of paints, or it can be applied in advance. Because of costs related with the uses and elimination of coatings, the shorter wavelength light sources are attractive to several uses.

Application

Mostly the laser heat treatment is used in automotive and machine tool industries who are responsible for process development as listed in table 4.1.

Industrysector	Component	Material
Automotive	Axel bearing seat	AISI 1035 steel
Automotive	Blanking die	Tool steel
Automotive	Engine valve	Alloy steel
Automotive	Gear teeth	Steel
Automotive	Shaft	Steel
Automotive	Piston ring	Steel
Automotive	Steering gear housing	Malleable cast iron
Domestic good	Typewriter interposer	AISI 1035 steel
Machine tools	Cutting edge	Steel
Machinery	Gear teeth	AISI 1060! low alloy steel
Machinery	Mandrel	Martensitic stainless steel
Machinery	Press brake tools	Steel
Machinery	Tool bed	Cast iron

Power generation	Turbine blade edge	Grey cast iron
Railway	Diesel enaine cylinder	Cast iron

Table 41 Some industrial uses of laser transformation reinforcing (Anon., n.d.)

CHAPTER 5

Machining Processes

INTRODUCTION

Machining is any of different processes in which a bit of material is cut into a coveted last shape and size by a controlled material-removal procedure. Machining is the wide term used to depict expulsion of material from a workpiece, it covers a few processes, which we as a general separate into the accompanying classifications:

- Cutting, by and large including single-point or multipoint cutting devices, each with an unmistakably characterized geometry.
- Abrasive processes, such as granulating.
- Nontraditional machining processes, using electrical, chemical, and ideal use of energy.

It is essential to see machining, as well as all manufacturing operations, as a system comprising of the workpiece, the apparatus and the machine. The discussions in this chapter cover introductions on subjects like mechanics and shear bending in machining, and heating in machining.

The customary machining incorporates turning, processing, penetrating, and crushing. It likewise incorporates computer applications which are being bolstered by the machining technologies. The non-conventional machining incorporates primers on the themes like ECM, EDM, AFM, USM.

In spite of the fact that machining is most normally utilized with metal products, machining is likewise done with materials such as wood, plastic, ceramic and composite materials.

MICRO ELECTRO DISCHARGE MACHINING (MEDM)

Diamond Micro Machining (DMM)

The demand to reduce the cost with greater precision and performance needs have directed a significant progress in advanced CNC diamond turning and grinding machines. Tendency to manufacture many strategic products for long time in which the tolerance are decreasing by a factor of 3 every year, on critical dimensions as highlighted in USA report.

For tighter tolerance and better surface integrities for costly components, the use of diamond cutting tool is increased. Ultra accurate cutting tools need to be tough and sharp and to have greater thermal properties in order to sustain their size and shape while cutting. Benefits offered by diamond include:

- Crystalline structure, which allows very sharp cutting edges to be produced,
- High thermal conductivity, the highest of any materials at room temperature,
- Ability to maintain high strength at high temperatures,
- High elastic and shear moduli, which reduce deformation during machining.

Jesse Ramsden, FRS in 1779 pointed in his earliest studies of diamond machining initiate to date describe the diamond turning. He machined a screw harned and tempered steel, with a diamond pointed tool, for use in his linear dividing engine for precision scale making. Diamond is, however, chemically react with ferrous materials at high temperatures, and is usually improper for the machining of steels and nickel alloys. This is due to very high wear rate of the diamond which results in nonviable tool costs.

In recent times, nonferrous metals like aluminum & copper are machined with diamond turning because in these materials it is difficult to obtain a mirror surface by conventional machining methods. Main reason is that these materials are relatively soft and abrasives produce scratches the finished surface and additionally incapable to produce high levels of flatness at the edges of the machined surface. However, diamond grinding can be successfully used for the machining of brittle materials like glasses and ceramics. The ability to control accurately the cutting tool position comparative to the work piece is an important benefit offered by advanced CNC diamond turning and grinding machines. This allows them to produce components that are very precise and accurate. (Jr, 2011) Conversely, the relative lapping and polishing. This makes it very hard to achieve accurate control of the tool's route for shapes other than simple geometric forms. Diamond machining is therefore proving to be a cost effective process for the manufacturing of intricate shaped components that have high precision requirements for form and / or surface texture. (Golam Kibria, 2017)

Diamond micromachining is specially used in optical and electronic industries. The processes are proficient of simultaneously achieving high profile precision, good surface finish, and low sub surface damage in brittle materials needed, for example, for semiconductors, magnetic read-write heads, and optical components.

Machining Principle

Very fine cuts and tiny chips can be produced by single-point diamond turning and ultra-precision diamond grinding.

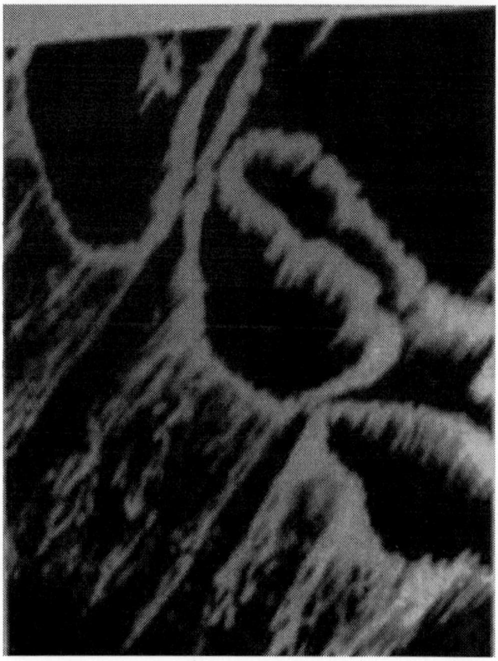

Figure 5.1 Scanning electron micrograph of electroplated copper, diamond turned at a depth of around 1 nm. (Company, 2004)

Figure. Illustrates a scanning electron micrograph of electroplated copper cut by a sharp diamond on an ultra-precision machine tool. The un-deformed chip thickness is around 1 nm. Because of the very fine chip width produced by the micro-cutting methods, the chip- making model for turning is different from that for grinding, moving from intense shear to micro extrusion as shown in Figure.

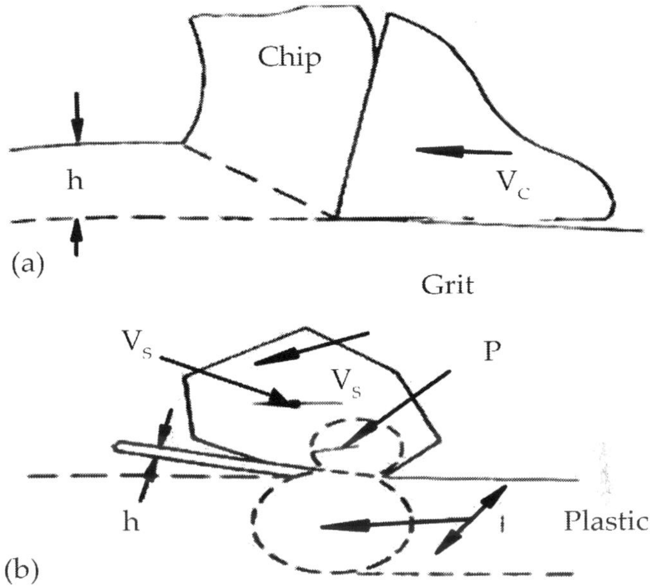

Figure 52 (a) Cutting concentrated shear model. (b) Fine grinding micro extrusion model (Black, 2015)

Impurities (inclusions) in the material, grain boundaries of polycrystalline materials, and homogeneities are the important features of materials for diamond micromachining. These can be a source of small vibrations of the cutting tool, resulting in wear in surface finish. Another factor affecting the quality of surface finish as well as consistency of form is the high coefficient of expansion combined with low thermal conductivity of some plastics which are diamond turned. These thermal effects are, to some extent, reduced when cutting with a diamond tool due to its sharp cutting edge, low coefficient of friction, and high thermal conductivity which transports the heat away. The academically peak-to-valley surface roughness which can be attained by diamond turning using a round-nosed cutting tool is narrow to:

Where ,

Rt stands for Theoretical peak to surface roughness (mm),

f stands for Feed-rate per revolution of the work-spindle (mm. rev -1),

Tr stands for Tool nose radius (mm).

Though, this equation overlooks any of the errors inherent in machine, and a more accurate equation takes account of the asynchronous error motion of the machine in the direction normal to the component surface. The actual peak-to-valley surface roughness now becomes:

Where

Esyn is the asynchronous error motion (mm) in the direction normal to the machined surface.

Brittle Materials

The need for the micromachining of hard and brittle materials has led to important progresses in machine tool technology. It is now possible to produce plastically distorted chips, when machining fragile materials, if the depth of cut is sufficiently small. This process is-to-exists when cutting brittle materials at small load machining is important for the cost-effective production of high-performance optical and innovative ceramic components, with very low levels of subsurface impairment (micro-cracking). This increases their performance and strength expressively and eliminates, or reduces, the need for post polishing.

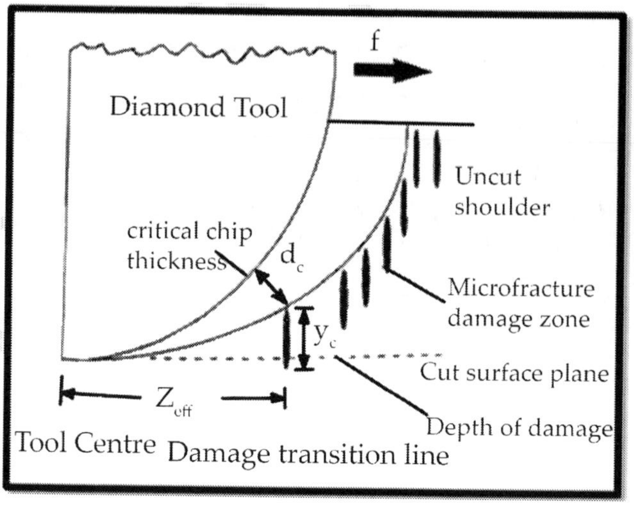

Figure 53 Cutting prototypical for the brittle/ductile system diamond turning of brittle materials. (Bralla, 2007)

The transition from ductile to brittle fracture has been generally reported and is usually described as the "critical depth of cut." This is usually small (i.e. 0.1 to 0.3 μm), as is the related feed rate, and this results in relatively slow material removal rates. However it is a cheap technique for producing high quality spherical and non-spherical optical surfaces, without the need for polishing.

Figure. Demonstrates the machining model for turning or fly cutting a brittle material, in which the depth of cut and critical chip thickness (dc) are shown. The position of the critical chip thickness is reliant on feed. For example, it is located on the way to the upper edge of the shoulder when ne feed-rates (f) are used and the micro-fracture damage zone is removed during machining. In this case subsurface damage does not extend into the cut surface. However, if the feed-rate increases, the critical chip thickness moves down towards the cut surface and this results in the micro-fracture damage penetrating into the final cut surface. In micromachining it is usually significant to confirm that these cracks do not occur by eradicating the material in a ductile mode. (Bralla, 2007)

The process requires careful selection of the machining factors in order to maximize the material removal rate while keeping high surface and subsurface integrity. It also demands high-precision, high rigidity machine tools with smooth motions.

Rates of Micromachining for Relevant Materials

Ductile mode machining is mandatory when machining mirror like surfaces in hard and brittle materials. Conversely, in order to achieve this condition the actual depths of cut required, to avoid crack generation, can be on the order of 0.1 to 0.01 of those used for cutting "mirror" surface in metal.

Diamond Turning

Gerchman and McLain initial work on the machining of germanium in which they diamond- turned germanium to a surface roughness of 5 to 6 nm These were sphere-shaped surfaces, 50 mm in diameter, for which the elimination rate was given in term of 2.5 μm per revolution of the work piece together with a 25 μm depth cut. More newly Shore has reported that removal rates the order of 2 to 4 mm3 per minute has been attained when diamond turning germanium optics of 100-mm diameter. The tool life (stated as the useful cutting distance of the too) when generating optical surfaces (<1 nm Ra) at these removal rates was in excess of 12 kilometers. When machining silicon at parallel removal rates, as with germanium, tool life was found to be less than 8 kilometers. 1 nm Ra is the order of surface finish quality in Micro-EDM. Tool life was greater, when machining zinc sulphide, being in surplus of 20 kilometers, although the surface quality was worse, with a roughness value of 3.6 nm Ra. (Golam Kibria, 2017)

Diamond Grinding

Continuous machining by a single point tool on a large area of brittle material can cause major problems if it needs to change the cutting tool when partially through a cut. A grinding wheel, however, has infinite cutting points (grits) produce a higher machining rate. Diamond micro-grinding therefore be estimated to improve the commercial viability the ductile mode machining of brittle materials.

While grinding is a multipoint process that relies on mechanical actions, it has been that chemical effects also play a substantial role in material removal rates when using micron -size abrasive grits to grind glasses in a ductile mode. A comparatively soft hydrated layer is formed on the glass surface the chemical reaction between the coolant and the glass.

For the elastic mode grinding of optical glasses, the material elimination rates of 0.75 to 1.55 mm3 per minute, when standardized for a 100-mm diameter optical component. This value was grown when producing surface rough-nesses of 1 to 3 nm Ra, which are convenient to what can be accomplished by the polishing process. A possible technique to achieve higher removal rates, when ductile grinding, is to uses very high grinding wheel speeds. These should, academically, decrease the un-deformed chip width, and thus the cutting force per grit, resulting in more ductile flow coupled with less strength degradation.

Accuracy and Dimensional Control of DMM

Diamond micromachining is used to produce either:

- Minor work piece features, by means of tools with cutting

features below 100 μm,

- Sub - μm or Nano metric tolerances and/or surface finishes on macro-components.

Very sharp-edged diamond tools have been used in the manufacturing of ultrafine optical gratings with a precision of 1 nm, and gratings with 1-nm resolution can now be obtained for use on ultra- precision machine tools. The most exact diamond turning and grinding machines currently available are capable of achieving geometric accuracies of size and profile on the order of 100 nm for dimensions of 250 mm. Surfaces of 0.8 nm Ra have also been diamond-machined on a number of materials, including germanium. Diamond micromachining demands tremendously smooth movements, particularly between the spindle and the tool. In order to accomplish this, hydrostatic oil and air bearings are generally required for the spindles and guide ways. Other severe requests required from the machine tool are:

- Very high loop toughness between the tool and work piece,
- The capability to apply and maintain very small depths of cut, as low as a few nm in some cases,
- Little thermal drift, and
- The capability to operate at uniform feed-rates over a wide range.
- Other aspects to be considered during the design stage include:
- The type of coolant and its application, filtration and temperature control,
- Work-holding methods.

Application of DMM

Over the last 30 years the number of applications for diamond turning of a wide range of materials, and the diamond grind-in£ of brittle materials (e.g. glasses and ceramics) has in-leased significantly following the technological breakthrough in direct CNC machining in the so –called "ductile regime," is free from retained brittle fracture damage. Wide range of common nonferrous materials can be machined by diamond may be as a single crystal or compacted polycrystalline form.

Diamond Turning

Single crystal diamond turning has been used to machine microgrooves 2.5 μm wide by 1.6 μm deep in copper. The slop of groves were produced with a surface finish to 10 nm Rmax, and the application was the fabrication of lens master discs for the molding of high efficiency grating lenses. Diamond turning is being used increasingly as a high-precision, high-production rate process for a wide range of products including:

- Spherical and aspherical molds for plastic opthalmic lenses and for medical instrumentation and micro-laser optical disc/ CD players.
- A wide range of reflecting optics components. For example, aluminum scanner mirrors, space communication and high-power machining laser optics, and aluminum substrates for glancing incidence mirrors for X-ray telescopes.
- Infrared hybrid lenses for thermal imaging systems. Typical materials include germanium, zinc sulphide, zinc selenide, and silicon.
- Aluminum alloy automotive pistons which are machined, in the cold state, to complex profiles with tolerance on the order of 3 to 10 μm.
- Aluminum alloy substrate drums for photocopying machines.

Diamond Grinding

There are, however, numerous new applications where components are required to have lower mass, higher hardness and wear resistance, improved chemical inertness, and higher strength and fatigue life, often while working at higher temperatures than before.

The worldwide research and development, ceramic and intermetallic materials are ready to be used in gas turbines, pumps, computer peripherals, piston engines, and many other engineering products on a much wider scale. For many of these applications grinding, in the ductile mode, is necessary in order to retain material integrity through the minimization of subsurface damage and micro-cracking, which reduce the strength and fatigue life of ceramic components. (K. G. Swift, 2013) Diamond micromachining technology for the efficient manufacture of opto-electronics devices is clearly of critical importance in ensuring its progress covering broadband light-wave communication, high-density optical memories, and optical parallel signal processing. Although process development has concentrated on VSLI technologies such as lithography, there is clearly scope for applying ductile mode grinding, slitting, and trenching techniques for the efficient manufacture of monolithic integrated optics components. These techniques are analogous to the ultra-precision grinding of magnetic memory disk file sliders (or flying heads), but now in the ductile regime. The introduction of free abrasives into fixed abrasive processing can be shown to improve surface finish and productivity. Several hundred components can be machined at one set- up to submicron tolerances, producing low-energy-loss contacting optical surfaces (<2 nm Ra and zero surface micro-cracks) lending themselves to kinematic design for submicron assembly, with large consequent savings in assembly and test labor costs.

ULTRASONIC MICRO MACHINING (USMM)

Principle of USMM

Micro ultrasonic machining (micro USM), is one of the efficient material removal processes especially suitable for the micromachining of hard and brittle materials. The principle of micro USM is shown in Figure. In micro USM workpiece which is placed on the workpiece table vibrates at ultrasonic frequency (40 KHz). Abrasive slurry is injected on the top of the workpiece. There is a rotating tool which hits the abrasive particles in the slurry which in turn hit the workpiece and chip away the material from it. The vibrations given to the workpiece aid in refreshing the slurry so that fresh abrasive particles are in contact with the workpiece and also in removing the debris from the tool workpiece gap. (Benedict, 1987)

The abrasive slurry acts as lubricating agent as well as coolant in reducing the frictional heat generated due to the movement of the abrasive particles on the workpiece and heat generated by the vibrations due to the transducer. The slurry also collects the debris from the machined area. In general micro USM is carried out with water as the medium due to its properties of excellent coolant, easy removal of debris from machining zone due to low viscosity, low cost and easy availability.

Particles affect both part and tool: material removal takes place at the workpiece; wear occurs on both the tool and particles. The process is therefore characterized by removal rate on the workpiece, tool wear, and abrasive wear. Ultrasonic machining can deal efficiently with brittle materials. The tools should be made from materials that can resist wear: they should be either ductile (aluminum alloys, steel, titanium alloys, nickel alloys) or extremely hard (diamond). The abrasive particles have to be harder than the workpiece material: aluminum oxide, silicon carbide, boron carbide, and diamond are used.

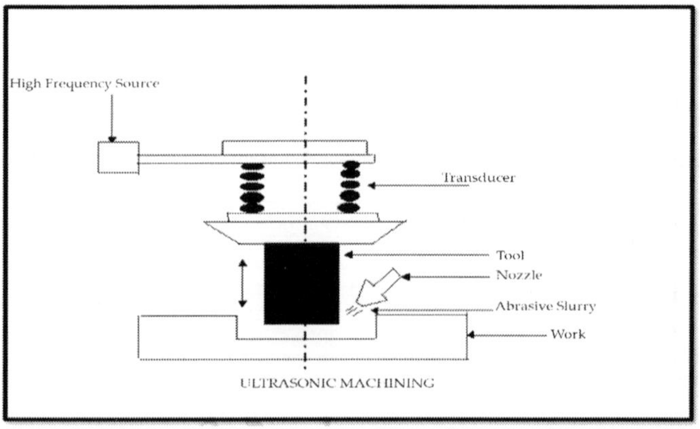

Figure 5.4 Illustration of Ultrasonic machining process (Anon., n.d.)

In many conventional machining processes like grinding, milling and broaching processes oil has been successfully used as cutting fluid. These oils can be used either as straight oils, which are pure petroleum based oils or emulsifiers which are water based oils. Use of straight oils have excellent lubricating properties and are used especially for machining process involving low speeds, low clearance requiring high quality surface finish. These oils have more viscosity and good lubricating properties than water and causes less tool wear. Hence, it may be prudent to use oil based slurry in micro USM.

Working

A micro USM system based on the design concept of "vibration on work piece" is schematically shown in Figure. A micro tool, attached to a mandrel rested on V- shaped block is rotated by a DC motor and is free to move in X, Y and Z directions with six degrees of freedom. Micro tools with different diameters are prepared by Wire Electrical Discharge Grinding (WEDG). The micro tool is sensitive to elastic bending, vibration, and breakage. Therefore, the contact force between tool and workpiece needs to be controlled and limited to a certain level during machining.

This is achieved by implementing a close-loop control strategy with force feedback. Key system components such as electronic balance and three-axis stage have high resolution (0.1 mg and 25 nm, respectively) to meet the demand of accuracy in micro machining.

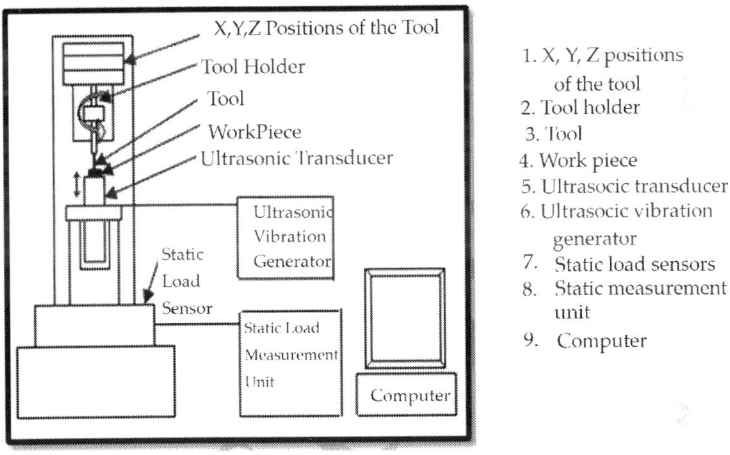

Figure 5.5 illustratiion of Micro USM (Anon., n.d.)

Material of Micro USM

Work piece Material

The USM process is able to machine any material, but is more efficient on brittle materials. Hard materials like stainless steel, glass, ceramics, carbide, quatz and semi-conductors are machined by this process. It has been efficiently applied to machine glass, ceramics, precision minerals stones, tungsten.

Abrasive Materials

The actual cutting tools are the abrasive particles. Their characteristics and dimensions have to be adapted to the material to be machined and to the specific application intended.

The hardness of the abrasive particles has to be higher than that of the workpiece material. For example, silicon carbide can machine glass, graphite, silicon, aluminium oxide, or precious stones; boron carbide has to be used for harder materials such as silicon carbide and silicon nitride. Diamond is the only abrasive able to machine even harder materials like diamond. For ease of use and owing to cost, boron carbide is often chosen for machining every material except diamond, and diamond is used only when boron carbide is insufficiently hard. Wear of particles is a crucial factor, since removal rate depends on particle diameter. Grains can be worn rapidly. (Black, 2015)

The abrasive slurry contains fine abrasive grains. The grains are usually boron carbide, aluminum oxide, or silicon carbide ranging in grain size from 100 for roughing to 1000 for finishing. It is used to microchip or erode the work piece surface and it is also used to carry debris away from the cutting area.

Tool Materials

Tool material should be tough and ductile. Low carbon steels and stainless steels give good performance. Tools are usually 25 mm long; its size is equal to the hole size minus twice the size of abrasives. Mass of tool should be minimum possible so that it does not absorb the ultrasonic energy. Two main considerations arise in the selection of the tool material; fabrication and cost. Tool wear during ultrasonic machining also has to be taken into account.

Piezoelectric Transducer

Piezoelectric transducers utilize crystals like quartz whose dimensions alter when being subjected to electrostatic fields. The charge is directionally proportional to the applied voltage. To obtain high amplitude vibrations the length of the crystal must be matched to the frequency of the generator which produces resonant conditions.

Material Removal Rate (MRR)

In machining with fine sized grains, the MRR increases with the increase in the abrasive particles due to increase in the number of particles involved in the machining. MRR is more when machining with the water as oil is more viscous than water hampers the process of debris removal in the processing of machining thus accounting for less MRR.

Using medium sized particles, MRR increases with the increase in the concentration as there are cutting edges for a given volume of abrasive slurry. The MRR is more when machined with oil compared to water as shows the comparison of MRR between water based and oil based slurry with respect to the varying abrasive slurry concentrations for the particle size of 1-3.

When machining with coarser grains the MRR is more when machined in aqueous medium compared to oil. As the grains become coarser the grain boundaries try to interlock reducing the number if cutting edges. Oil possessing more viscous property interlocks these grains strongly compared to water thus contributing to less MRR.

Accuracy and Tolerance in USMM

In general, needed accuracy entails both roughing and finishing, since quality can seldom be obtained in a single operation. Roughing is performed with large grains (20 to 120 μm) to give sufficient removal rate; finishing is achieved through 0.2 to 10 μm) to obtain the desired quality. Drilling of very small holes is performed in a single operation. Tool wear is of major consideration for accuracy, since it affects both tool geometry and dimension.

Accuracy strongly depends on the machining mode (Fig 4.). In sinking, it is the result of tool initial accuracy tool wear abrasive dimensions and working parameters. The lateral gap between tool and part is found bet one and two times more than the abrasive main diameter, frontal gap is a little larger, due to amplitude of vibrations. Fluctuations of gap are smaller for smaller grains. In general when drilling, the use of roughing (40 μm) and finishing (5 μm) can provide ±10 μm accuracy. When finer grains, about 1 μm or less, are used, accuracy can be better than ±5 μm. this includes the tool accuracy. It is difficult to provide estimate of accuracy for very small holes (10 to 200 μm), because of difficulties with tool accuracy. In contouring, accuracy can be better, since tool imperfections can be compensated by 3-D movements. (Black, 2015)

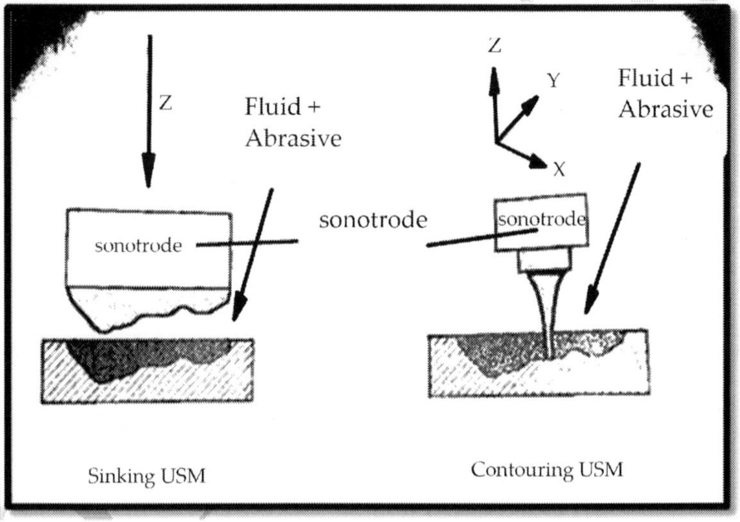

Figure 5.6 Ultrasonic machining variants (Anon., n.d.)

In ultrasonic micro machining, since very fine grains are used, ± 5 μm or better accuracy can be obtained when tools are conventionally made. A higher accuracy can be achieved by using specially manufactured tools (e.g. the tool is produced on the USM machine, by use of wire EDM)

Surface Roughness

When machining with finer grains the surface, finer grains having constant cutting edges hit the work piece repeatedly. Further the debris is added in the process of material removal increasing the frictional heat making the surface rougher with the increase in the concentration. However oil acting as coolant reduces the surface roughness to some extent. Thus producing good surface finish compared to machining in water medium.

Using medium sized abrasives the surface roughness, with the increase in the concentration the particle size decreases in the process of machining, indicating more number of particles to absorb the heat generated during the process which eventually reduces the surface roughness. However surface roughness for oil is less compared to water since water acting as coolant absorbs the generated heat.

When machining with coarser size grains, as the concentration increases there are coarser particles hitting the work surface thus making it rougher. Since oil acts as better coolant than water thus giving better surface finish.

Advantages

- Machining any materials regardless of their conductivity
- USM apply to machining semi-conductor such as silicon, germanium etc.
- USM is suitable to precise machining brittle material.
- Can drill circular or non-circular holes in very hard materials
- Less stress because of its non-thermal characteristics.
- It can be used machine hard, brittle, fragile and nonconductive material.
- It is burr less and distortion less processes.

Disadvantages

- USM has low material removal rate.
- Tool wears fast in micro USM.
- Machining area and depth is restraint in micro USM.
- It is difficult to drill deep holes, as slurry movement is restricted.

Applications

- It is mainly used for
- Drilling
- Grinding
- Profiling
- Coining
- piercing of dies
- Welding operations on all materials which can be treated suitably by abrasives.
- USM can be used to cut industrial diamonds
- USM is used for grinding Quartz, Glass, and ceramics.

- Application of USMM can be found in electronics, aerospace, biomedicine, and surgery.

MICRO ELECTRO DISCHARGE MACHINING (MEDM)

Micro-electro-discharge machining is also known as micro-EDM and electro discharge micromachining, has been invented in the past 30 years from the nonconventional manufacturing technique of electro-discharge machining (EDM) usually known as spark erosion.

EDM has been used as a production tool for half century, in 1967 when Kurafuji and Masuzawa succeeded to machine 6 μm diameter holes though GTi 10 cemented carbide 50 μm thick, there has been a intensive effort to improve the micro-machining rates of several materials, without loss of accuracy and to increase the excellent surface finish and dimensional control already connected with the EDM techniques.

Many companies in Switzerland and japan are producing μ-EDM equipment commercially, known as center of excellence in micro-technology and precision engineering. μ-EDM is used to machine different types of miniature and micro parts from electrically conductive materials like metals, alloys, sintered metal, cemented carbides, ceramics, and silicon. It is also used to produce molds and dies that are used in manufacturing of other micro parts both conductive and nonconductive materials.

Basic Principles of Micro Electro Discharge Machining

Micro electro discharge machining (Micro-EDM) is modified form of EDM, which is normally used to manufacture micro and tiny parts and components by using the conventional electro discharge machining principles. Similar to conventional EDM, material is removed by a series of quickly repeated electric spark discharges between the tool electrode and the work piece in Micro-EDM. Major difference between Micro-EDM and conventional EDM is type of pulse generator, the resolution of X, Y & Z component and the tool size. Micro- EDM uses very small pulse produced by generator in the pulse period of a few micro seconds or Nano seconds. Due to this Micro-EDM uses low discharge energies to eliminate little volume of material. Machining ability of Micro-EDM on all type of conductive and semi-conductive materials with surface precision irrespective of material hardness is the major factor which makes the Micro-EDM very important. It is favored chiefly for the machining of difficult-to-cut material due to its high proficiency and accuracy. (Company, 2004)

Insignificant volumetric material elimination of Micro-EDM provides considerable prospects for manufacturing of micro-dies and micro-structure such as micro holes, micro slot, and micro gears etc. The usage of Micro-EDM has many returns in micro-parts; the foremost advantage is that it can machine intricate shapes into any conductive material with very low forces. The forces are very small as the tool and the work piece do not come into contact throughout the machining process. This property offers advantages to both the tool and the work piece. For example, in EDM, a very thin tool can be used because it will not be bent by the machining force. The other advantages of Micro-EDM include low set-up cost, high aspect ratio, improved precision and large design choice. In addition, EDM does not make direct contact between the tool electrode and workpiece material, later removing mechanical stress, chatter and vibration problems during machining. Therefore, trusting on the above advantages, Micro-EDM is very effective to machine any kind of holes such as small diameter holes down to 10 μm and blind holes with 20 aspect ratio.

Although Micro-EDM is a very effective process in micro hole machining and having many advantages, it has also some limitations. One of them is it is a slow machining process; the other is that while the work piece electrode is being machined, the tool electrode also wears at a rather substantial rate. This tool-wear leads to shape imprecisions. One more drawback is the development of a heat affected layer on the machined surface. Since it is difficult to remove all the molten part of the work piece, a thin layer of molten material remains on the work piece surface, which re-solidifies during cooling. (Golam Kibria, 2017)

Reverse polarization of tool electrode is another significant factor in Micro-EDM, which results the long pulse duration in conventional EDM, commonly anode is the charged tool electrode to increase the material rate and decrease the electrode damage. Micro EDM works at short pulse duration, here tool electrode is mostly charged as cathode. Micro-EDM can be applied on different machining processes; it can be

Micro-ED drilling, Micro-ED milling, Micro-ED die sinking, Micro-ED contouring, Micro-ED dressing, and Micro or wire electrical discharge grinding (Micro-WEDG). All these methods integrated in a today's sophisticated micro-EDM machines.

Figure. Displays the picture of the Micro-EDM setup. The main components of the Micro-EDM setup are the machine tool itself, the pulse generator, the CCD camera associated with the monitor and the microscope for examination. A more detailed image of the Micro-EDM machine tool is displayed in Figure The Figure. Displays different parts and components of the Micro-EDM machine tool.

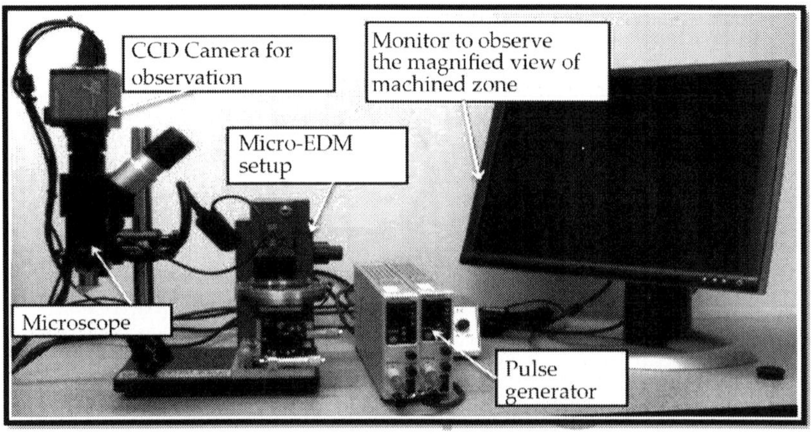

Figure 5.7 Micro EDM Arrangement (Company, 2004)

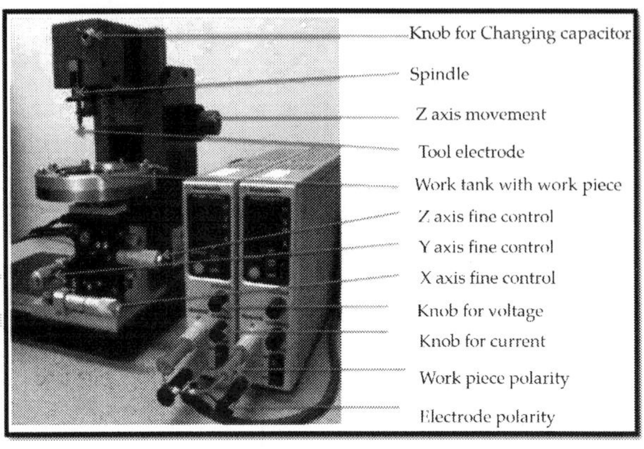

Figure 5.8 Detailed image of the Micro- EDM machine tool (Company, 2004)

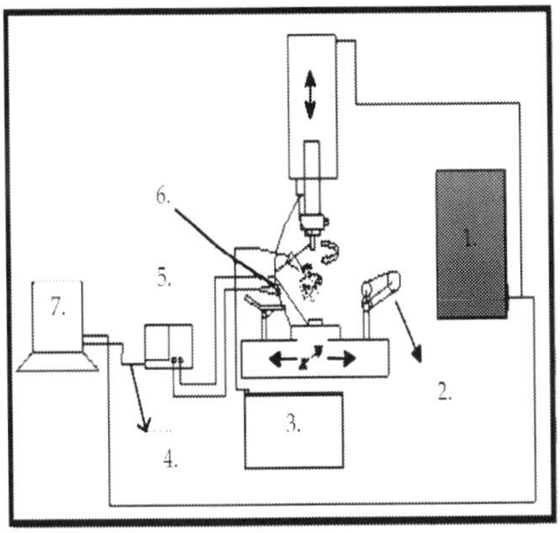

Figure 5.9 Schematic drawing of Micro-EDM system (Company, 2004)

A machine made four axis movements Micro-EDM machine as it is shown in Figure 2. Is used for micro hole machining. The machine has a pulse generator. This generator is able to produce pulses from 50 ns to 2 μs with electrical current peak values up to 50 A. Power supply can vary voltage levels from 50 V to 250 V, but this is limited up to 100 V for the used MicroEDM machine. In addition to the micro erosion generator, the micro-EDM machine is also composed of the following parts; control unit panel. Micro-electro discharge machining (MicroEDM) is a process of removing electrically conductive materials by rapid, repetitive spark discharges from pulse generator with dielectric flowing between tool and work piece. (R.L. Timings, 2009)

Electro-discharge machining (EDM) is well suited for micro machining high-strength conductive materials since neither mechanical contact nor cutting is necessary. Materials such as stainless steel and tungsten carbide are machined easily with electro-discharge and negligible cutting forces are applied to the tool or workpiece. The main disadvantage of electrical discharge is that during each discharge some material is removed from the tool. The melting temperature, conductivity, and change of yield stress due to temperature determine the wear rate of the tool and workpiece. (Golam Kibria, 2017)

Micro-electro discharge machining process is a widely used micro fabrication technique to produce micro-parts and components needed in the micro-mechatronic systems and industrial applications. Micro-hole fabrication is a primary task for this thesis because micro-hole is the most simple and widely used micro products that can be manufactured by using Micro-EDM.

WIRE ELECTRO-DISCHARGE GRINDING (WEDG)

The WEDG process, illustrated in Figure 3, is similar to turning on a lathe. A simple RC circuit generates pulses that produce electrical discharges between the workpiece (anode) and a ϕ 100 μm brass wire (cathode).

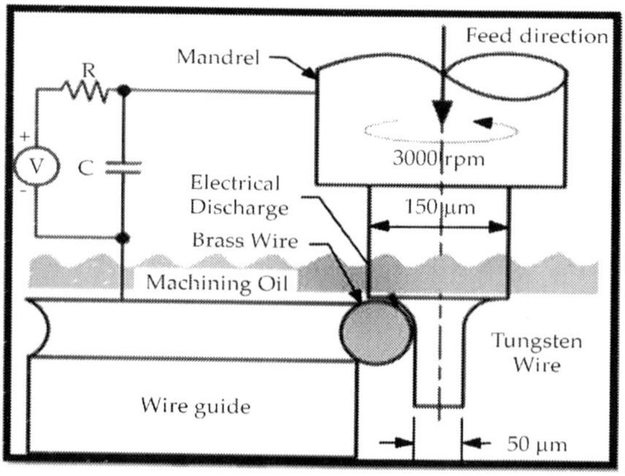

Figure 5.10 Wire Electro-Discharge Grinding (Anon., n.d.)

The discharges occur across a small gap (~ 2μm) filled with dielectric oil. The workpiece is held vertically in a mandrel that rotates at 3000 RPM, and its position is slowly fed in the z-direction. The wire is supported on a wire guide, and its position is controlled in the x- and y-directions. Each electrical discharge erodes material from the workpiece and the brass wire. To prevent discharges from worn regions of the brass wire, the wire travels at 340 μm/s, and is fed around a reel and take-up system as illustrated in Figure . (Davis, 1995)

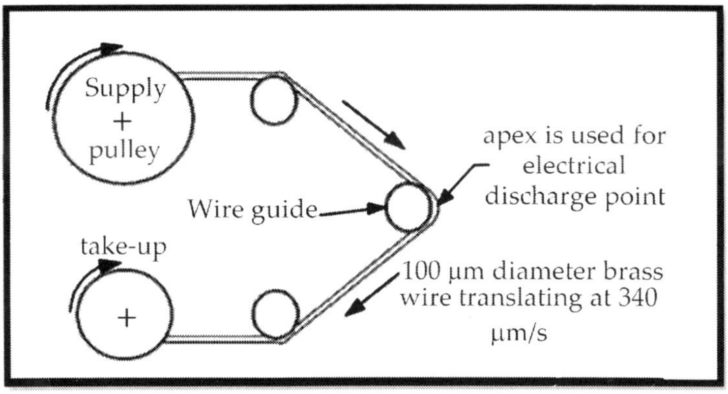

Figure 5.11 Traveling Wire in WEDG (Black, 2015)

Operation	Geometry	Process Conditions
Step 1: Flat End		V = 100 V C = 1 Fh = 30 µm/s F1 = 2 µm/s
Step 2: Rough Cut		V =110 V C = 1 Fh = 30 µm/s F1 = 2 µm/s
Step 3: Finish Cut		V = 70-80 V C = 4 Fh = 20 µm/s F1 = 30 µm/s

Figure 5.12 Typical Steps and Conditions for WEDG (Black, 2015)

A micro shaft is usually produced in three consecutive steps as illustrated in Fig 4. In the first step, the workpiece is positioned above the traveling wire, and the end of the shaft is machined by feeding the wire/guide in the x-direction. The second step is to rough cut the shaft and reduce the diameter of the stock material by feeding the workpiece in the z direction. A high material removal rate (MRR) is achieved during the rough cut by increasing the energy of each discharge, which depends upon the energy stored by the capacitor as given in Equation (1). The final step is to finish cut the shaft. The voltage and capacitance are reduced to achieve improved form and surface finish. Although the multi-step process is based on the premise that improved precision is obtained by reducing the capacitance and voltage, a numerical relation for straightness or roughness is not available.

Substantial effort has concentrated on the precision of holes or cavities machined by micro-EDM using cylindrical electrodes made by WEDG. Masuzawa et al. [7,8] used a vibroscanning method to measure holes drilled by micro EDM, and Yu et al. [9,10] developed the uniform wear method to reduce inaccuracy arising from electrode wear when micro-machining cavities. Yu et al. [11] later studied the influence of current, voltage, layer depth, and feed on the material removal rate, electrode wear ratio, and gap during contour milling with a cylindrical electrode. (Black, 2015)

Rates of Micromachining for Relevant Materials

The rate of machining depend on the discharge energy, which for the execution of fine surface finish is kept low (<10-7 J per pulse) in μ-EDM.

A number of techniques have been consumed to try to increase the material removal rate such as:

- Switching a larger capacitance into the RC circuit, thus producing a large discharge energy, pielectric flushing,
- Jump action and vibrofeeding of the tool, premachining holes to allow debris to flow away from the electrode, and
- Use of controlled pulse generators in WEDM.

Accuracy and Dimensional Control

The accurateness and dimensional control of parts made by μ-EDM differs with types of machines used.

Maximum hole diameter	=	300 μm
Minimum hole diameter	=	10 μm
Wire material	=	Tungsten
Wire diameter	=	30 μm
Part and part material	=	Injection die and inbox
Surface finish	=	0.15 μm

Maximum dimensional variation = ± 1

Machining Time = 47 minutes

Application of -WEDM

The applications of -EDM are many and various. :

- Micro-pins & shafts
- Micro pipes
- Inkjet nozzles
- High aspects ratio holes
- High aspects ratio slots
- Square cornered cavities
- Dies
- Micro-gear wheels
- Special orifice

Applications of μ-EDM

Micro-EDG appears to be the least common. However, it has been stated that 60-mm long channels, 900 μm to deep and 60 μm wide with blind ends have been machined into both sides of a stainless steel plate to form part of a micro-reactor. Such micro-reactor structures are also used in mixing chambers, heat exchangers, and pumping systems and in such materials as titanium diboride.

A round flat disc rotating at high speed, removes material along its circumference as work piece moved in the opposite direction of to the rotary movement as shown in fig. A cylindrical electrode has been used to machine micro-grippers for high-precision assembly of RF circuits by pick-and-place units.

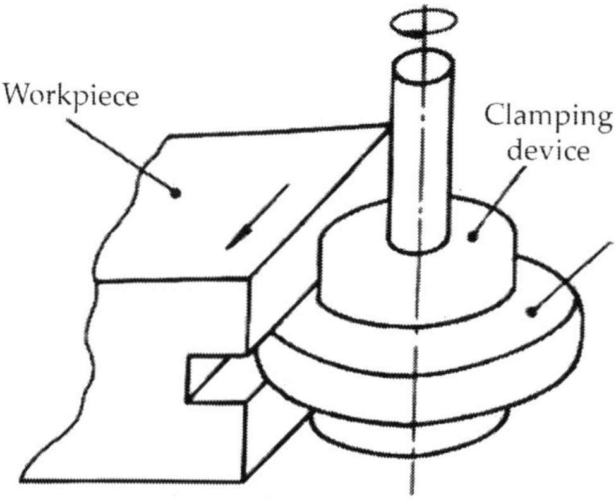

Figure 5.13 Micro-EDG. (Black, 2015)

CHAPTER 6

Reverse Engineering

INTRODUCTION

In This chapter techniques and innovations for computerizing reverse engineering (RE) through advanced imaging and PC vision are described. The chapter start this part with a meaning of RE regarding producing computer-aided design (CAD) design from existing data and parts. This chapter utilizes the term computer-aided reverse engineering (CARE) to depict this procedure. With this definition, this chapter shows a concise review of the conventional way to deal with RE utilizing coordinate measuring machines (CMMs). At that point, the chapter starts the primary concentration, where it investigates advanced imaging and PC vision as an other option to CMMs. This investigation starts with information securing, where we pre-dispatched laser-based range scanners as a promising methodology. For these scanners, the contrasts in information determination and checking rates and contrast those to CMM execution are clarified and highlighted. Next, it exhibits a handling pipeline for making CAD models utilizing these scanners. This preparing incorporates tasks such as view enrollment, surface coordination, patch remaking, abstract fitting, noise elimination, and different capacities. This chapter discloses these points to enable you to understand their significance in planning a RE imaging system and the effect that different parameters have on demonstrating precision.

COMPUTER-AIDED REVERSE ENGINEERING

Reverse engineering (RE) has numerous implications to a wide range of individuals. As this chapter starts, we initially concentrate our consideration on the particular implying that we plan to address and then stretch out our thoughtfulness regarding the robotization of RE, which we call computer-aided reverse engineering (CARE). To this end, this part talks about CARE as a general and then investigates 3-D laser filtering inside and out as a developing CARE innovation from the computer vision group. The chapter describes the approaches and systems associated with computer vision scanning and notes particular difficulties for the CARE implementation. (Bessant, n.d.)

REVERSE ENGINEERING DOES NOT INCLUDES

Each train of engineering has an alternate definition for RE. Computer engineers and computer researchers, for instance, allude to RE when they talk about determining the algorithmic usefulness of a software bundle when they have no earlier information of the programming design. Developers and engineers endeavor to create practical block charts of the software product through interaction with the interface and to get high level code depictions from crude machine code. This product definition is not the extent of our RE dialog. Another case of RE that may be familiar but additionally outside the extent of this chapter–concerns uncovering the inward workings of a machine to make sense of what makes it tick. This type of RE is likewise a systems level approach where a designer disassembles the thing important to build up an understanding of the utilitarian relationship of parts or to pick up knowledge into the sorts of materials used to create the segments. As with programming RE, the objective is to build up an higher state portrayal of a system without from the earlier information. These two cases are normal applications that utilize the term RE, yet we wish to emphasize that our meaning of

RE is not identified with these cases, but rather is identified with the range of computer-aided engineering (CAE).

COMPUTER-AIDED ENGINEERING

In between the l 1970s and 1980s, computer-aided design (CAD) which is a segment of Computer aided engineering began to reform engineering disciplines. The pinnacle of this insurgency happened in 1990 with the design of the Boeing 777; the whole flying machine was designed and preassembled through a virtual CAD) reenactment. As per Boeing, the initial 777 to roll out of the production shed in 1994 was simply hundredths of an inch about the thickness of a playing card within arrangement. This accuracy contrasts with the half-inch arrangements regular with most air ship parts before that time a change of a few requests of extent. This insurgency in innovation has proceeded into the 1990s and today with the emergence and development of computer-aided manufacturing (CAM). computer-aided manufacturing is the mechanization of the manufacturing procedure itself–beyond simply the design process– where machines such as computerized numerically controlled (CNC) factories permit exact creation of articles specifically from CAD portrayals. With CAM, a designer and developer can quickly move from a calculated CAD portrayal to a true substantial product. We may utilize the term forward engineering—in an offhanded way—to portray this sort of design, and the term CAE to depict the robotization of forward engineering through CAD and CAM advances. (Benedict, 1987)

COMPUTER-AIDED REVERSE ENGINEERING

Computer aided reverse engineering through Computer aided design and Computer aided manufacturing advancements is the computerization of engineering and production, where a design formalizes ideas through computer displaying and then manufactures those models into certifiable products. CARE streams the other way. CARE makes a computer abstract / model of a product through measurements of the product, as it exists in its present reality. In this specific situation, we characterize CARE as the inversion of CAE or the capacity to produce a CAD abstract from a genuine real world product. We outline this stream in Figure. The disc brake shows up on the left half of this figure and its CAD demonstrate shows up on the privilege. A laser-based range scanner was used to make the CAD show. This model is metrically precise up to a couple of millimeters of the original brake disc. By similarity, one may think about this capacity as a 3-D fax. Similarly as a copy (fax) machine changes over a hand-composed record into advanced shape, we have changed over an unmistakable produc into a computer display. This figure represents our meaning of CARE. To be more solid, we characterize CARE as far as the geometry and state of a product.

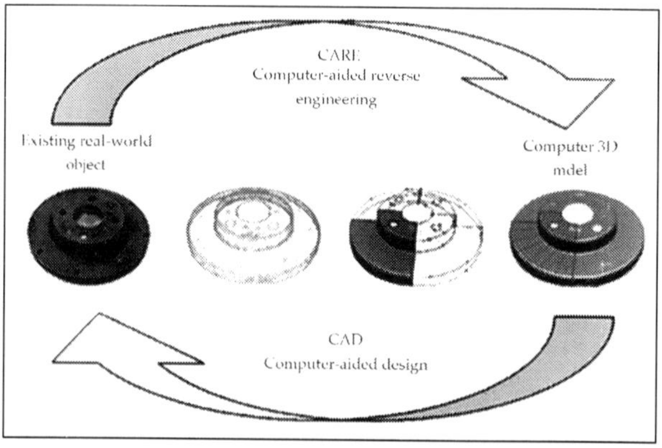

Figure 6.1 Computer-aided reverse engineering (CARE) process (Raja, 2008)

To understand the CARE steps, consider the steps appeared in Figure. From left-side to right-side. The initial phase in the CARE procedure is to make measurements at specific focuses along the surface of the object. Every place and point on the object is having a x, y, and z components describing its position in the 3-D space. For a given product, a CARE system will measure hundreds or even thousands of such focus points relying upon the design idea of the product and the kind of CARE system. The accumulation of these focus points is called as a point cloud; an illustration shows up in the next picture from the left in the Figure. In many cases, the point cloud is an adequate depiction of the product. In any case, more elevated amounts are conceivable as the two pictures on the right-side illustrate. The third picture from the left-side is a component portrayal of the product, in which the CARE system has recognized surface edges and crevices from the point cloud information. The last picture on the figure is a full and finish CAD portrayal of the product. For this portrayal, the CARE system utilizes the point cloud and the identified components to fit surfaces for displaying the whole geometry of the question. (Company, 2004)

For both modern and military applications, CARE offers many focal points to the engineering design, manufacturing creation, and field support of a section or part. For instance, CARE permits fast investigation and approval at the production line based on the first CAD designs. A production professional can rapidly assess resilience in respect to the CAD models. Such feedback data empowers more tight control over the manufacturing procedure or may help future redesigns of the segment. As another illustration, since CARE makes computerized models, the innovation is perfect for electronic scattering of information. A manufacturing setup in china can transmit as-constructed models of segments to a design focus in London or the other way around. Albeit numerous thousands of miles isolate these two setups, CARE empowers more proficient coordinated effort. The London group can transmit their CAD design to china by means of electronic mail, and the Chinese group can restore a CARE model of a created product. In the past, such electronic shipping was a restricted attempt where the physical model would require costly post based conveyance back to London. At long last, chronicled issues of field adjustments and out-of-production segments are another zone for which CARE systems offer guarantee.

Architects regularly experience circumstances where nontechnical faculty create novel and vital alterations of their designs however simply after they are delivered. Such field-operations work force are insufficiently prepared to coordinate their adjustments to the designers. These changes are normally impromptu with almost no documentation, such as CAD drawings. Therefore, engineers have little expectation in the event that they wish to catch these adjustments and join them into future designs. Nonetheless, the potential exists for even nontechnical faculty to send back a 3-D fax of the alteration to the specialist. Furthermore, engineers themselves confront a comparable issue when asked to adjust an out-of-production part for which CAD documentation does not exist or has been lost. The specialist has the repetitive task of creating a CAD drawing without a CARE system. This capacity is the guarantee of CARE to permit both specialized and nontechnical individuals to produce engineering quality CAD models of existing articles rapidly and consistently. (Raja, 2008)

In the event that a CARE system naturally catches the geometric structure of a product and stores the resulting shape and topology data as a CAD display, the following inquiry may be the way to accomplish this objective. By what method can a CARE system itself measure the geometry of a product? In industry, the most widely recognized answer is through a coordinate measurement machine (CMM). In spite of the fact that coordinate measurement machine are not the computer vision arrangement that this part investigates, they are the beginning stage as an industry standard for a talk of CARE. Thus, in the following area, we quickly survey the capacities of a CMM and then contrast those abilities with the computer vision approach that laser scanners offer.

COMPUTER VISION AND REVERSE ENGINEERING

The figure demonstrates a measurement of the brake utilizing vernier calipers. Calipers are a typical impromptu way to deal for RE. These measurement gadgets permit engineers and mechanics to decide precise widths, lengths, and other dimensions of products. This way to deal with RE is a manual procedure that requires much exertion for convoluted products and surfaces. CMM innovation is the main push to robotize the RE procedure. Before CMM and presumably still well known for most basic tasks, specialists and mechanical engineers have utilized measurement calipers. In case of disc brake, one could utilize calipers to measure the distances across the different openings and curves that contain the basic details of the brake, as in Figure. At this point, from these measurements, we could physically lay out a computer model of the brake utilizing CAD primitives. For a basic product, this manual procedure of RE is direct, however as the multifaceted nature of the product shape increases, the basic CAD primitives such as planar and quadric surfaces are not any more reasonable. A freestyle surface, for instance, that is non-planar and non-quadratic does not permit itself promptly to portrayal with just calipers. Freestyle surfaces require extraordinary thought and consideration. Calipers are not down to earth for catching their nuances. (Raja, 2008)

As an option, CMMs first showed up in the mid-1960s and are a more functional implies for portraying and reviewing freestyle surfaces.

A CMM comprises of a measuring probe upheld on three commonly opposite axis x, y, and z; every axis has a worked in reference standard. Figure gives an applied perspective of a CMM. The measuring probe permits exact measurements along every axis with respect to the standard. In this manner, a CMM produces 3-D arrange focuses as the measuring probe moves over a surface. Engineering design Developers may run a CMM in a manual mode where they move the measuring probe around an object and gather design measurements, or they

Figure 6.2 Measuring the disk brake using a caliper (Raja, 2008)

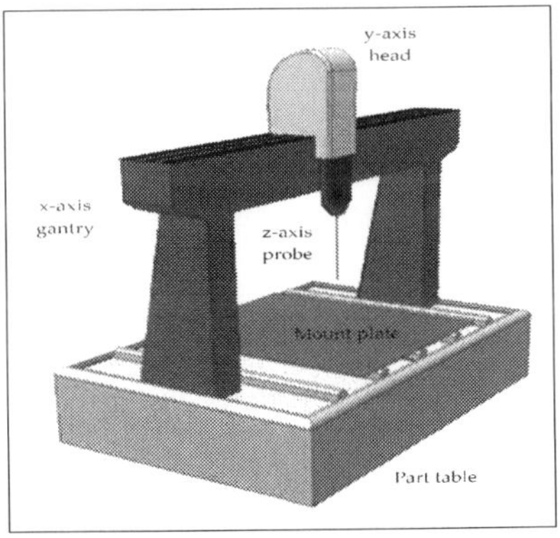

Figure 6.3 nceptual view of a CMM that illustrates the major components of most systems (Raja, 2008)

Mechanize the measuring probe to move over the object's surface automatically. This strategy which is described later is more pertinent to the CARE definition under consideration. Diverse CMM manufacturers offer distinctive plans to enable manufacturers to design the way that the measuring probe will take after. The more progressed CMM systems enable engineers to transfer a CAD model of the product and then the CMM utilizes this model for the measuring probe arranging procedure. The CMM will investigate the CAD model to distinguish basic focus points and areas such as irregularity creases or tight spaces. Tight spaces are a state of emphasis on the grounds that the measuring probe must interact with or be near the point at first glance where a measurement is done. On the off chance that the measuring probe can't achieve the indicated due expansion limitations i.e., the fact is too far or restricted space i.e., the measuring probe is too vast, at that point the system can't measure at that area. The measuring probe must touch or be close to the measurement area. This requirement is the inspiration for computer vision ways to deal with CARE since computer vision offers a more standoff measurement.

Active Illumination 3-D Stereo

Computer vision endeavors to create designs through sensors such as camcorders or other imaging systems. To get 3-D details, the most surely understood technique that presumably strikes a chord is stereo vision, in which two cameras work in a way like human eyes. Awkwardly, albeit stereo-based imaging has turned out to be valuable in such areas as route planning and hindrance evasion for Mars Rover missions, passive stereo does not offer the precision that mechanical RE applications need. The significant downside of stereovision is the need to build up connection between a point in one camera's picture and a similar point in the second camera's picture. If we can set up this correspondence, we can build a precise stereo profundity guide, or range picture, yet correspondence is a subtle issue. The term passive stereo is imperative since it demonstrates that the cameras in the system don't depend on dynamic types of enlightenment, with the exception of surrounding light. One approach to conquer the correspondence issue is to control illumination in a shrewd way and accordingly streamline the correspondence look. Such strategies are known as dynamic brightening stereo (or all the more essentially dynamic stereo) where we supplant one of the stereo cameras with an all-around characterized light source. A laser is normally the light wellspring of decision. (Raja, 2008)

Dynamic stereo utilizing laser-based light permits more exact profundity measurements than conventional passive brightening stereo. (We utilize the term profundity measurement as an other option to a 3-D point measurement from CMMs. Now, the inconspicuous subtlety between these two terms is not imperative, but rather they are tradable on the grounds that a profundity measurement is a 3-D point measurement.) Although passive stereo is not feasible for a CARE system, dynamic stereo offers potential. To understand dynamic stereo, we ought to consider three mainstream approaches to enlightenment control. These real classes are

(a) consistent wave modulation

(b) time-of-flight estimation

(c) organized light triangulation.

Consistent wave systems measure profundity by utilizing a cognizant light source, i.e., a laser, and measure phase contrasts of the light wave as it goes to and from the product. Figure demonstrates an illustration. The separation of a product part from the camera is relative to the phase distinction between the transmitted wave and the returning wave.

The precision of these systems relates with the exactness in measuring the phase contrasts and the wavelength of the laser. Time-of-flight techniques work also to radar where the system measures the time required for a laser beat to go to and come back from a product. Figure outlines such a system. Very exact profundity measurements require accuracy hardware to measure speed-of-light time changes. As a third choice for dynamic stereo, organized light techniques figure profundity through geometric triangulation. The camera, the product, and the light source shape the triangulation geometry. See Figure This arrangement is like passive stereo geometry, aside from that we have supplanted one of the cameras with a light source. The exactness of these strategies is a component of the camera determination, geometric measurements, and brightening exactness, yet the essential parameter for controlling and increasing precision is the camera resolution. System geometry and enlightenment are not as basic. Along these lines, organized light systems offer an answer more pragmatic than passive stereo in accomplishing the accuracy important for a RE system. In a later area, we will investigate organized light strategies more inside and out, however first we try to thoroughly analyze the advantages of dynamic stereo to CMM. (Raja, 2008)

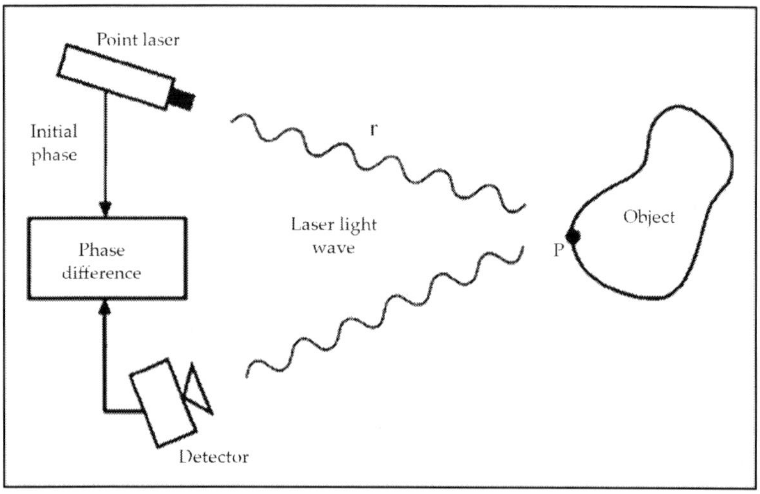

Figure 6.4 Active stereo example of a continuous wave system. The laser travels as a light wave from the laser source to the object and back to the detector. As a wave, the laser light undergoes a phase change as it travels. (Raja, 2008)

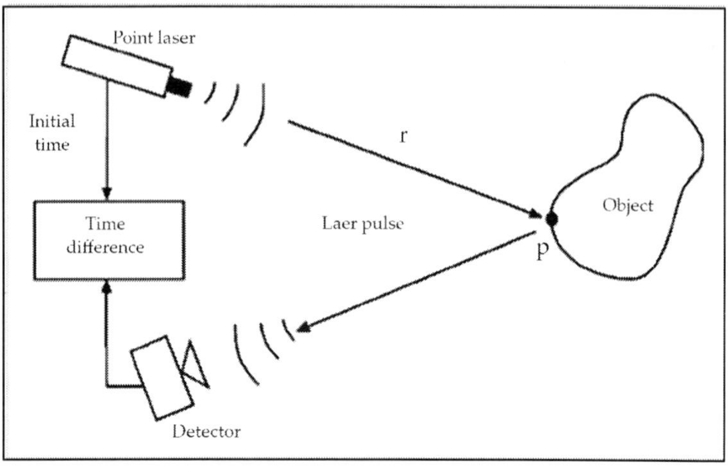

Figure 6.5*Active stereo example of a time-of-flight system. A point laser emits a pulse that is reflected from the object of interest. (Raja, 2008)*

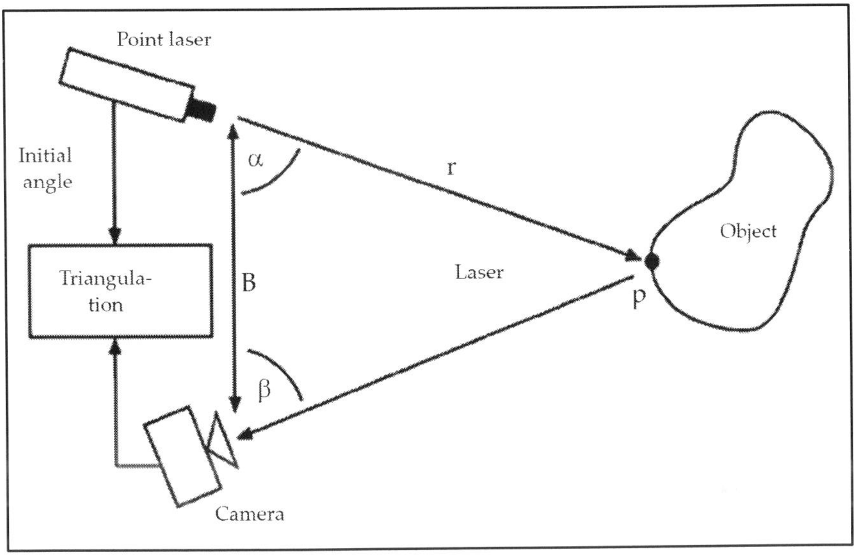

Figure 6.6 Active stereo example of a structured-light system. This example shows a triangulation-based approach where B, β, and α are typically known through system calibration. (Raja, 2008)

Benefits and Drawbacks

In the previous parts, we examined CMMs and dynamic stereo as techniques for CARE. Presently we investigate the advantages and downsides of these systems in cost, precision, speed, and unwavering quality. Since we are not talking about particular manufacturers and on the grounds that dynamic stereo is a rising innovation for CARE, we confine ourselves to a subjective correlation at a more hypothetical level, as opposed to a quantitative examination at a showcasing level.

To start, we take note of that CMMs are for the most part mechanical systems. In spite of the fact that they have computer control and utilize best in class gadgets for making measurements, they are bound by physical standards where a measurement probe should physically move around a product near its surfaces. Cameras and range scanners, then again, are strong state sensors that take measurements from a certain distance. Subsequently they require less mechanical accuracy in development. The precision of a CMM is based specifically on the mechanical control of the measuring probe, however the exactness of a dynamic stereo system depends more on its strong state sensors. A quick comparison is to consider the distinction between simple watches with riggings and springs and computerized watches with transistors and quartz gems. With solid state systems, this similarity infers different purposes of intrigue such as the advantages of less moving parts, the benefits of speed, and the ease of programming re-configurability. (Chee Kai Chua, 2014)

As noted already, CMMs must contact the product of importance to measure its geometry. By contrast, computer vision systems are non contact in that a camera makes measurements at some standoff separation. (We ought to clear up that some business CMM systems claim to be non contact gadgets, yet they don't necessarily permit standoff measurements. Despite everything they require a measurement probe to be close to the point of measurement, just not touching it.) The standoff idea of computer vision systems offers two favorable circumstances. The first is that they keep away from damage to either the product or the measuring probe. As a measuring probe checks and moves over a surface, it might scratch that surface if the client neglects to handle the measuring probe legitimately or if the surface itself is excessively weak or profoundly moldable. Also, if the temperature of the surface is excessively hot or excessively cold, the warmth exchange could harm the measuring probe. These situations are not a noteworthy worry with computer vision systems. A commonplace laser-based range scanner can measure points on a product at a separation of 10 to 20 centimeters, if not more.

The second preferred standpoint to the standoff capacity is the disentanglement of path arranging. With a touch measuring probe, a CMM (or a client) should precisely choose a measurement way that legitimately covers the surfaces of a product yet that abstains from wedging the test into tight spaces. The CMM must utilize a way that covers the product surface and yet complies with the physical limitations forced by connection between the product and the measuring probe. Commonly, a CAD model of the object is important to help path arranging by filling in as an indicator of the product shape. Path arranging for a computer vision system is less prohibitive. A camera or laser range scanner can float around the product and gather measurements similarly as a man taking photos of the product may do. Some arranging is important to guarantee that we measure the whole product, yet the system does not need to check physical constraints as with a CMM measuring probe. Once more, a relationship is illustrative. Consider two men one of whom is wearing a blindfold. The two men are asked to depict a product. The man with no blindfold will basically get the product, turn it around a couple of times to acquire distinctive perspectives, and then report his depiction. Since he can't see, the second man must be somewhat more cautious as he utilizes his fingers to test around the question and investigate its general structure by touch. His path arranging is clearly more confused than that of the first man. This similarity needs adequate advancement, as with most summing up analogies. It unintentionally infers that CMMs and touch measuring probes are less exact than computer vision strategies, however the similarity conveys the substance of path arranging with each approach. The previous talk highlights the principle favorable benefits of computer vision systems for CARE, however, their weaknesses frustrate their useful use over CMMs. As a newcomer to RE, computer vision arrangements are a youthful technology that has not achieved the business development that CMMs have accomplished. Right now, CMM innovation has huge market infiltration. The current capital interest in CMM tilts industry toward redesigning as opposed to replacing CMMs. The familiar quote–if it ain't broke, don't settle it–gives assurance to this imperviousness to embracing computer vision systems. Additionally, business computer vision systems for CARE are recently starting to rise up out of the re-seek group and many research questions are as yet uncertain. Among these inquiries are the manner by which to enhance precision. Past market entrance, the most critical drawback of computer vision systems is exactness. Ultra-high-exactness CMMs work in the 1–2 micron range, and more direct (as far as cost) CMMs in the 10–20 micron extend. Computer vision techniques can-not contend—as of yet—with these levels where most systems work in the about millimeter scope of exactness. The pattern, in any case, demonstrates that CMMs have leveled. Just a couple of microns distinction in precision bringing about more than a $100 thousand increase in cost. Then again, computer vision inquire about shows that more noteworthy exactness is yet to accompany more exact lasers and higher resolution imaging sensors. With this expectation, the following segment will dive further into the most well-known computer vision approach an organized light system, and later we explore a considerable lot of the examination issues in this approach (Company, 2004).

STRUCTURED-LIGHT RANGE IMAGING

We start this segment by first laying out three classes of organized light solutions to 3-D go imaging. At this point, we select one of these classifications, the sheet-of-light technique, and talk about it in more detail. This area looks to clarify the way in which a computer vision system gathers 3-D point data.

Source Illumination Categories

In spite of the fact that we have specified a few ways to deal with range imaging in a past segment, the most encouraging for CARE are the organized light methodologies. Remember that these are triangulation techniques, in which the geometry of the laser, camera, and product are set to recuperate the separation from the camera to the object. This separation, or range esteem, is an element of three parameters: the base-line distance between the light source and the camera, the view angle of the camera, and the light throwing angle of the light source. Review the geometry in Figure We have the accompanying condition for the range esteem, r:

$$r = F(\alpha, B, i, j, \ldots)$$

Where F is a function, possibly nonlinear, α is the illumination angle, and B is the baseline distance between the camera and the light source. The variables i and j represent the horizontal and vertical positions, respectively, of a pixel in the camera image. (A pixel is the fundamental picture element from a digital camera. We commonly represent images as a matrix where a pixel is an element of the matrix. We classify the quality of a digital camera in terms of the number of pixels. For example, a 1-megapixel camera usually has 960 rows × 1280 columns for its pixel matrix.) This pixel is the brightest spot where the light source falls on the imaging plane of the camera. The point on the image directly correlates with the range, i.e., distance, of the object from the laser. The sequence in Figure demonstrates how the pixel location changes with increasing range. Due to the view disparity between the light source, as the object moves further away from the light source, in this case a point laser, the projection of the laser onto the image plane from where it strikes the object moves as well. When the object is closest to the laser, the projection is near the left edge of the image plane, but when the object is farthest, Figure the projection is near the right edge. (Benedict, 1987)

In Figure, we have used a point laser for illumination, but other types of light sources are possible. The three most common sources of illumination are single-point lasers, sheet-of-light lasers, and coded-light patterns. As already discussed, Figure shows the point laser. If we do not move the object, we can scan the laser in a raster fashion across the scene to obtain a range measurement for every pixel. If the image has M × N pixels (i.e., 0 i < M – 1, 0 j < N – 1), then we require M × N measurements. The second type of illumination is the sheet-of-light laser shown in Figure. It projects a plane of light onto the object so that this plane intersects the object and forms a line. That line becomes a projection onto the image plane of the camera. This method is a direct extension of the point laser approach, except that the line represents a collection of simultaneous point sources. Thus, one can readily see that a sheet-of-light system allows one to increase scanning speed over a point laser with almost no increase in effort. The sheet-of-light method requires only M measurements for an M × N image. The third method is the coded-light source. This method illustrated in Figure is

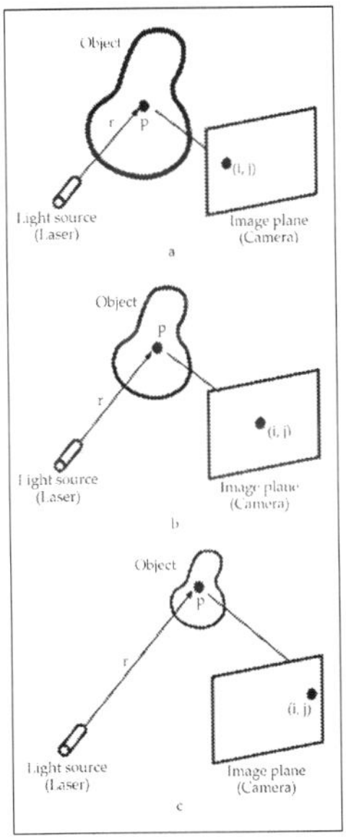

Figure 6.7 Demonstration of how a structured-light system determines distance. (a) The object is close to the laser source. (b) The object has been moved

a slight distance further away. (c) The object is at an even greater distance. Note the change where the laser (Raja, 2008)

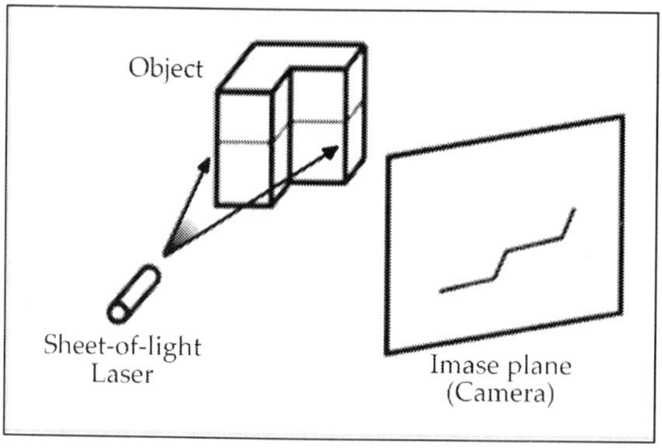

Figure 6.8 Sheet-of-light range scanner. The laser source fans out to create a plane that intersects the object of interest. The resulting line of light on the object projects as a line onto the imaging plane of the camera. (Raja, 2008)

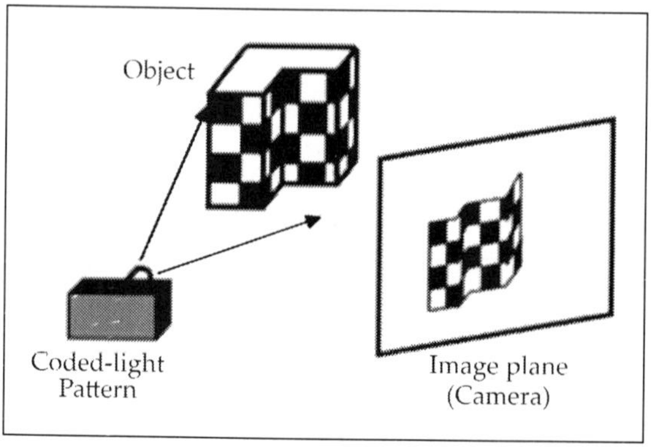

Figure 6.9 Coded-light range scanner. The light source is usually a projector with a screen pattern placed over the lens. The pattern projects onto the object and subsequently projects onto the image plane of the camera. (Raja, 2008)

Entirely different from both of the earlier two strategies. Ordinarily, this technique does not utilize a laser source yet rather utilizes a slide projector with patterns put in the slide merry go round. The system extends these patterns as coded light onto the product, as the figure shows. The examples take diverse structures however one technique utilizes a checkerboard. The range determination of this strategy is an element of the quantity of examples utilized. In this manner, a M × N picture would require R measurements where R = log(M) in the event that we look for a determination like a sheet-of-light output.

The essential contrast between these three enlightenment techniques is the speed with which one can gather information. Precision changes little among them. At first, one may feel that the coded-light strategy offers the fastest checking speeds, yet practically speaking, the components important to trade and show patterns onto a product are a bottleneck. The sheet-of-light technique fits the fastest implementation. Business sheet-of-light scanners are accessible that can produce 10,000 information points every second, and this number is developing as innovation technologies progress. Such speeds are past the abilities of a mechanical CMM measuring probe that would need to move around the product at a high rate. These information rates are the novel preferred standpoint that computer vision laser scanners–specifically sheet-of-light scanners–bring to CARE. In the following area, we investigate these scanners. (Prasad, n.d.)

Sheet-of-light Range Imaging

Sheet-of-light scanners offer the best speed advantage in gathering 3-D information contrasted with other laser scanners and hence are the most appropriate for CARE uses. Their basic operation is such that the laser extends a line onto the product of intrigue, and triangulation among this line, the laser, and a camera yield 3-D measurements. The photograph in Figure is a case of a laser line anticipating onto an arrangement of items, and Figure 2.10b is a case of a system, the Ranger Scanner.

In the figure, the items are laying on a transport belt. This belt is one technique for getting a full output of a product. With the Ranger Scanner, as with other sheet-of-light scanners, a solitary measurement brings about a solitary line–a profile– of information. This profile is the 3-D information for points where the laser line falls on the object. An arrangement of such profiles over the product is important to shape a total 3-D depiction of the product. In this way the transport line moves the product under the laser line to check the profiles over the whole product. When we stack the subsequent profiles together, we have a 3-D model of the product. The transport produces profile cuts of the product. Another approach is to not move the product but give motion to scanner. Some sheet-of-light systems adopt this strategy where for the most part the camera and laser are attached on a wand. The client swipes this wand around the product of intrigue. With the belt, the separating between the profiles is basically a component of the travel velocity of the belt and the acquisition rate of the scanner. With the wand, a more complex system for tracking is important to decide the inter profile relations. Business arrangements vary from magnetic position trackers to CMM probes. (Raja, 2008)

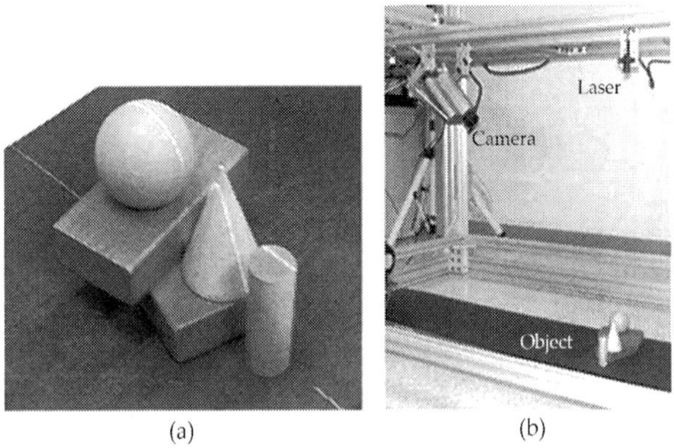

Figure 6.10 Examples of a sheet-of-light system. (a) The laser sheet casts a line on the objects under measurement. (b) This system is the IVP Ranger Scanner. (Raja, 2008)

As for precision, the geometry characterized in Figure and the determination of the camera represents the system. The most widely recognized game plan is for the view angle α to be in the 30–60° territory, and camera resolutions are in the range of 512 × 512 pixels or more. These pixel resolutions correspond with more than nine bits of range determination. As the view angle α increase toward 90° or decreases toward zero, the range exactness decreases. At these extraordinary review points, the camera has trouble recognizing the positional changes in the laser. Initially, let us consider a zero view angle. In the event that the product surfaces are planar and opposite to the laser plane, the camera won't see the laser on account of the slanted angle, however this angle yields the best projection for profundity changes. Furthermore, take the 90° angle. With this view, the balance camera is practically looking straight down the laser plane. The camera identifies the laser at a similar position paying little respect to profundity changes on the product's surface. This view offers the most exceedingly awful projection to the camera, yet it gives the camera the best perspectives of the laser reflection. Hence, the 30–60° territory adjusts these two issues of non-oblique survey and projection resolution. This exchange off is additionally an element of the baseline distance B between the camera and the laser. As in customary passive stereo, this separation ought to be of about an indistinguishable extent from the range augmentations of intrigue. (Raja, 2008)

The last issue we consider is impediment, which is a typical issue with computer vision systems. All the more particularly, the issue is called self-occlusion

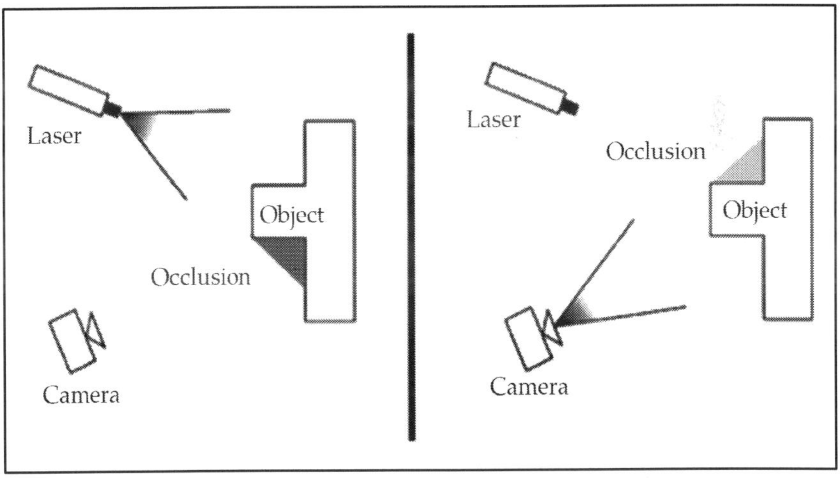

Figure 6.11 The occlusion problem with a sheet-of-light scanner. The object may occlude both the laser and the camera from viewing different surfaces on the object. (Raja, 2008)

Where a piece of a product impedes the review of another piece of a similar product. A spherical product shape is a straightforward yet not really clear illustration. At the point when a camera sees a spherical product shape, the camera can picture the front side, yet it can't image the reverse side. The front blocks the back. Shapes that are more complex than a circle have more complicated occlusions. Since the camera and the laser must view the protest, a sheet-of-light range scanner has two conceivable outcomes for impediments. As Figure represents, the camera and laser can't see similar surfaces of the product. The camera may picture a surface that self-impediment does not enable the laser to light up. Then again, the laser may light up a surface that self-impediment escapes the camera. An assortment of techniques are available to limit impediments, however we can't remove the issue, as the basic spherical shape case illustrates. In the following segment, we examine a pipeline for building CAD models utilizing laser range scanners. We particularly examine different perspectives as a vital answer for impediment. (Raja, 2008)

SCANNER PIPELINE

The aim of this subdivision is to plot the means that are fundamental for a laser range scanner to catch 3-D data of a real product and to make an interpretation of the data into a CAD description. We detail these means as a pipeline in Figure 1. Since large portions of these means are flow zones of dynamic research in computer vision, we display a diagram of the pipeline itself and identify the exploration challenges that face the practical deployment of a computer vision CARE scanner. Várady et al. (1997), Thompson et al. (1999), Guo (1997), and Campbell and Flynn (2001) also define pipelines like Figure

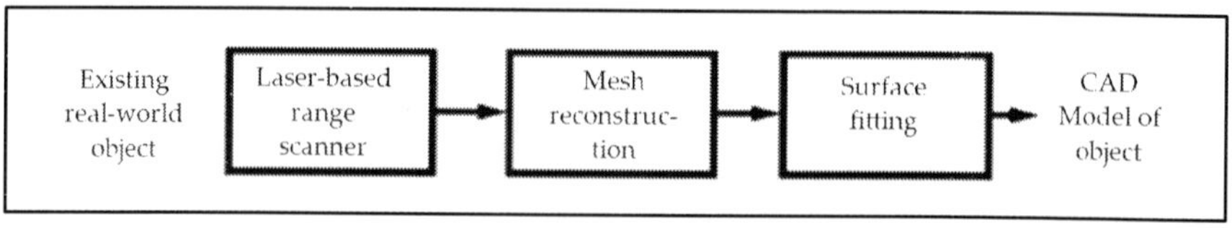

Figure 6.12 Block diagram of a CARE system based on a laser range scanner (Raja, 2008)

Data Collection

The initial step for a CARE scanner is data gathering. We have already covered this issue genuinely well in the past area on structured-light methods. Sheet-of-light coordinatess are most likely most appropriate for the CARE application. How-ever, we should discuss a couple of other vital themes, and we take note of that Suk and Bhandarkar (1992) provide additional purposes of discussion. (Raja, 2008)

The main subject we consider is calibration. Figure demonstrates how calibration allows changing a range picture into a 3-D point cloud.

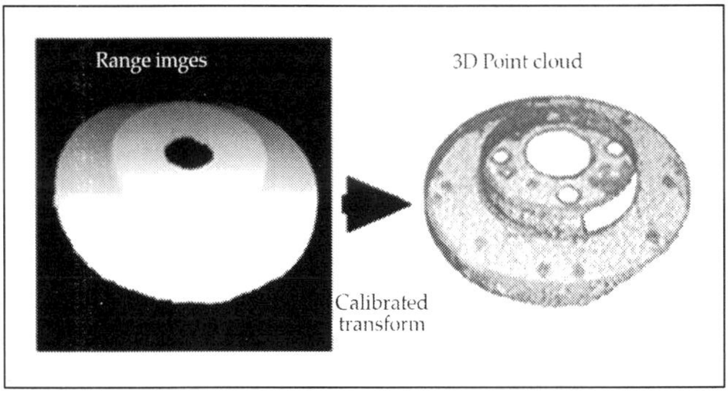

Figure 6.13 represents the distance between the brake and the sheet-of-light scanner. The darker pixels are farther away. If we calibrate the scanner, we can transform this range image into a 3-D point cloud, as shown on the right. The points (dots) exist in 3-D (Raja, 2008)

The term range image are using to some extent slackly without accurate definition. A range of image is similar to the traditional digital image that one might find with a digital camera. Taking a pixel (i,j), the distance to the object in the sight is the range (value of r) while the units of r proportional to the arrangement of the system which cannot be directly identified. Hence the raw data from the scanner is an image of pixel data; nevertheless we do not essentially identify the units for neither r nor the relative of pixels to those units. Calibration controls the connection between (r, i, j) and some space coordinate system (x, y, z) in units, for example, meters or inches. With alignment, we can characterize a change T that changes over a range picture into a 3-D point cloud, Where p is a point in the cloud. The one on the privilege is a screen shot of a 3-D point cloud. The point cloud is difficult to picture on a 2-D sheet of paper, yet utilizing a 3-D viewer, we would be competent to explore in and around the focuses that the screen shot shows statically. (Raja, 2008)

After adjustment, our next critical issue is see enlistment. In the former area, we expressed that impediments are troubles that prompt truant information. We noticed that restricted to paralyze this issue is to move the range scanner around the question acquire diverse review edges. Figure represents how at least two perspectives are fundamental to gather estimations of the whole brake. On the off chance that the scanner is on one side of the brake, little helpings of the opposite side are blocked. The test with different perspectives, be that as it may, is that we should now enlist the progressive coordinate system. Look at that as a point p from one view has organizes (x, y, z) near to one world coordinate system, yet that comparable point in the other view has coordinates p' = (x', y', z') with respect to a moment arrange coordinates. Despite the fact that p and p' speak to physically a similar point on a surface, the diverse view estimations prompt distinctive coordinates system.

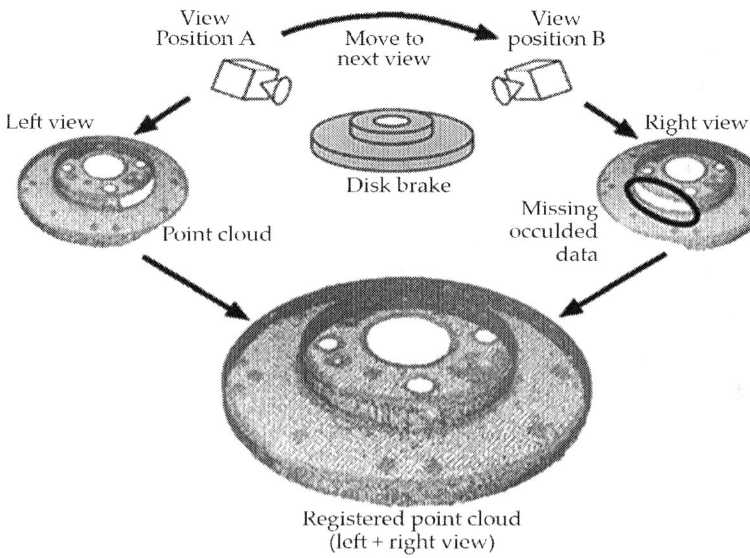

Figure 6.14 Multiple view scanning and registration overcome occlusions. Part of the brake is excluded in each of the views above. When these views are registered together, one view fills the occluded regions of the other view. (Raja, 2008)

Enlistment is the way toward recuperating the firm change that relates these two arrange coordinates. The condition for this change is

$$p = Rp' + t,$$

Where the grid R is a revolution and the vector t is an interpretation. The issue is that R and t are unclear. Enlistment algorithms recover these factors using the crude information from various perspectives of the question. Horn et al. (1989) introduce a shut shape answer for the recuperation itself. One measure is the measure of cover among the perspectives. If no cover exists, enrollment is for all intents and purposes inconceivable unless we force higher request requirements.

Some fractional cover is important to determine the enlistment. A CARE scanner using laser extending must characterize the measure of cover between sees, as the scanner gathers information around the product.

Mesh Reconstruction

When we have gathered different perspectives of a question and enrolled those perspectives to make a point cloud, the subsequent stage for the CARE scanner is to reproduce a work that models the surface of the product draw an obvious conclusion of the point cloud, in a manner of speaking. Without much idea, a great many people at first consider work recreation insignificant, and it most likely is insignificant for the human identity, yet programmed remaking by a PC is not all that simple. Hoppe et al. (1992) first tended to this theme for the general instance of a chaotic point cloud. From that point forward, numerous re-searchers have introduced their answers, and an overview showed up in Campbell and Flynn (2001). Figure 4 shows work reproduction. This figure is a zoom perspective of the hover brake to show the triangle work recouped from the point cloud. Triangles are a default standard as a surface primitive for an assortment of reasons, yet basically they rearrange PC representations since they have ensured convexity and in this way are helpful as a first-arrange guess of a product. (Later we will discuss surface fitting, which prompts higher request approximations. (Raja, 2008)

For recreation, examining characterizes the quantity of focuses in the point cloud and their relative separation to each other (i.e., their thickness) on the product. Examining is an exchange off between informational collection size and product scope. A PC would more be able to promptly store and control a littler informational index; however a bigger set all the more precisely catches the devotion of the question. Under testing prompts deficient and mistaken models; however oversampling has a tendency to overpower registering assets. Work remaking calculations must characterize their examining necessities, and clients must tailor their CARE applications in like manner. Also, characteristic in inspecting is estimation blunder. In spite of the fact that this mistake is an information obtaining issue, most work recreation calculations consider this issue to limit the impacts of such blunder. On the off chance that we have a 3-D point p on the surface of a question, measuring that point with a scanner unavoidably acquaints a mistake with such a degree, to the point that the watched point q has the going with connection:

Q = p + e

Where the vector e is the error. Work reconstruction algorithms endeavor to limit the effect of e. notwithstanding measurement error, a reconstruction algorithm must consider three elements. One is the topology of the object. Topology concerns the global structure of the object. For instance, a ball and a donut have diverse topologies. In any case, a block of wood and the ball have a similar topology. The topological gap among the donut is the distinction. A dialog of topology is beyond the extent of this part, but we need to stress that it is a vital thought for reconstruction. The second factor is the measure of cover between various perspectives of the object. Already, we have noted in Figure the need to gather various perspectives to conquer self-impediment. Typically, these numerous perspectives cover in their scope fundamentally to help enlistment for Equation, but cover is also valuable for limiting measurement error. A reconstruction algorithm requires a technique to handle numerous perspectives and their cover. The third factor is missing information, which prompts openings in the work. These gaps are not topological gaps as with the donut but rather crevices in the work like set patterns on a sheet of paper. Generally, specialists have created two classes of work reconstruction algorithms to deal with these elements of topology, cover, and missing information. These classifications are surface-based, for example, exhibited by Bernardini et al. (1999) and volumetric-based methods, for example, created by Curless and Levoy (1996).

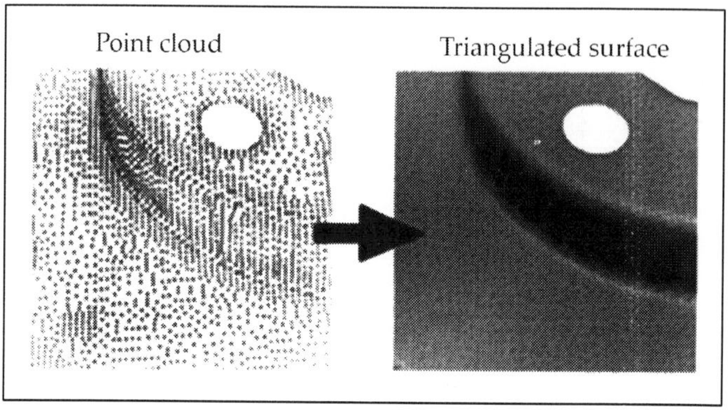

Figure 6.15 Reconstruction recovers the triangulated surface from the 3-D point cloud data (Raja, 2008)

Surface Fitting

When we have a first-order estimation of the object from work reproduction, the last phase of the pipeline is to create higher request depictions that are more suitable for CAD applications. Now and again, a triangle work itself is adequate as the last result of a CARE scanner. In different cases, parametric portrayals, for example, non-uniform sound B-splines (NURBS) are important. We broadly label this last stage as surface fitting, but by and by, it takes many structures and is profoundly reliant on the CARE application. Eck and Hoppe (1996) show a strategy for the recuperation of splines from extend checks, and their paper exhibits the difficulties that we layout. A parametric portrayal of a surface takes the accompanying structure: (Raja, 2008)

where u and v parameterize the surface and the parameters f, g, and h restore the x, y, and z facilitates, separately. For most intricate products, a solitary parameterized surface is not feasible and not commonsense for the whole question. Or maybe, we should isolate the question into surface fixes and afterward show each consequent fix with its own parameterized surface. In this manner, our first undertaking is to fragment the questions into suitable fixes and after that parameterize those patches separately. The significant test is recognizing suitable patches, especially; be-cause parameterized surfaces expect a rectangular structure. Rectangles don't fit the self-assertive limits that regularly happen with fix determination. Accordingly, an extra stride must trim the surface to these limits. With a fix distinguished, the fitting for recouping the parameters for f, g, and h is direct. Both immediate and iterative advancement arrangements are normal. Figure, is a straightforward delineation of surface fitting for the now natural brake. In this figure, we utilize highlight recognition of surface discontinuities to fill in as the fix limits and afterward fit NURBS to the subsequent patches. This figure finishes up this section and concludes the CARE methodology, initially appeared in Figure.

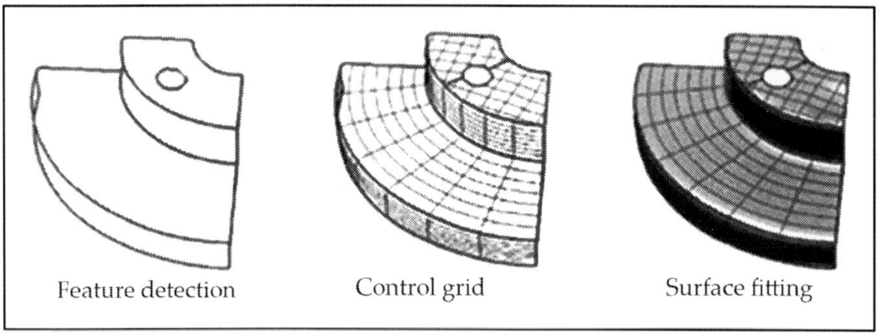

Figure 6.16 Typical surface fitting sequence. The leftmost image shows the feature detection to establish potential surface discontinuities. The next image to the right shows the control grid to support the surface fitting process. Finally, the rightmost image shows (Raja, 2008)

CHAPTER 7

3D Printing

INTRODUCTION

3D printing is also known as "Additive manufacturing" a manufacturing method. In this process the material is added in successive patterns instead of removing material to create the desire shape.

3D printing, otherwise called additive manufacturing, alludes to processes used to make a three-dimensional object in which layers of material are processed under computer control to make an object. Objects can be of any shape or geometry and are delivered utilizing computerized drawing information from a 3D model or another electronic information source such as an Additive Manufacturing File (AMF) document. In this manner, not at all like material expelled from a stock in the traditional machining process, 3D printing or AM fabricates a three-dimensional product from computer-aided design (CAD) model or AMF record by progressively including material layer by layer.

The expression "3D printing" initially alluded to a procedure that stores a binder material onto a powder bed with inkjet printer heads layer by layer. In the recent past, the term is being utilized as a part of mainstream vernacular to encompass a more extensive assortment of added substance manufacturing methods. Joined States and worldwide specialized standards utilize the official term additive manufacturing for this more extensive sense. ISO/ASTM52900-15 characterizes seven classes of AM processes inside its significance: cover flying, coordinated energy statement, material extrusion, material streaming, powder bed combination, sheet overlay and vat photopolymerization

THE HISTORY OF 3D PRINTING TECHNOLOGY:

3D printing is another innovation, the introduction of 3D printing was in 1984 because of Chuck Hull who imagined a procedure known as stereo lithography, in which layers are included by curing photopolymers with UV lasers, from that point onward, 1990 layer by layer innovation utilized each layer has 0.1mm profundity, in 1999 the main use in solution, in 2000 the initial segments of human, for example, ears, fingers was done, 2005 3D printing innovation ended up noticeably open source, in 2006 the primary SLS (specific laser sintering) machine wind up noticeably factor, in 2008 the principal self-replication printer which made the printer ready to print the lion's share of its own segments likewise at that year 3D innovation created to do a hard shapes and craftsmen for planners ,in 2009 Atom by iota printing were done which takes into account Bio3D printing, in 2011 the initial 3D printer Robotic Aircraft at that year the world's initial 3D-printed Car and it turned out to be economically accessible at the following year, at that year the principal gold and silver gems were finished utilizing 3D printer. (David Ian Wimpenny, 2016)

MAIN APPLICATION OF 3D PRINTING:

- Prototyping
- Specialized parts used in aerospace, military, biomedical engineering, dental
- Hobbies and household use
- Future applications specially in medical (body parts), buildings and cars

Software is used in 3D printing that slices the model into layers approximately 0.01 mm in thickness or less in some cases. Each layer is then sketched onto the build plate by the printer, once the shape is completed, the build plate is pull down and the next layer is added on top of the previous one. (David Ian Wimpenny, 2016)

Subtractive manufacturing is typical manufacturing technique in which mater in removed from the preformed block. Examples of subtractive manufacturing techniques are milling, cutting etc. lots of waste is created in this type of machining method because the material is cut off and can't be used for anything else and is simply gone to scrape. (Chee Kai Chua, 2014)

3D Printing removes such waste since the material is deposited in the location that it is required only, the remaining will be left out as empty space.

ADVANTAGES AND LIMITATIONS:

Greater flexibility and creativity is available in design process due to layer by layer production. No longer do designers have to design for manufacture, but as an alternative they can create a part that is lighter and durable by means of improved design. Parts can be entirely re-designed so that they are durable in the areas that they need to be and lighter overall.

3D Printing meaningfully speeds up the design and prototyping process. There is no difficulty with creating one part at a time, and changing the design each time it is produced. Parts can be created within hours. Transporting the design cycle down to a matter of days or weeks related to months. Also, since the price of 3D printers has reduced over the years, some 3D printers are now within economic reach of the ordinary consumer or small company.

The drawbacks of 3D printing in general include costly hardware and costly materials. This leads to costly parts, thus making it difficult if you were to compete with mass production. It also needs a CAD designer to generate what the customer has in mind, and can be costly if the part is very complex. (Chee Kai Chua, 2014)

Every type of production cannot be done by 3D Printing; though its innovation have helped to accelerate design and engineering more than before. 3D printer designer have the capability to create one kind of model of art, complex building and product designs and also make parts in space.

We are starting to see the influence of 3D printing many industries. There have been articles saying that 3D printing will bring about the next industrial revolution, by refunding a means of production back within reach of the designer or the consumer.

TYPES OF 3D PRINTING:

Main types of 3D printing

- Fused Deposition Modeling
- SLA – Stereo lithography
- SLS - Selective laser sintering

FDM – Fused Deposition Modeling

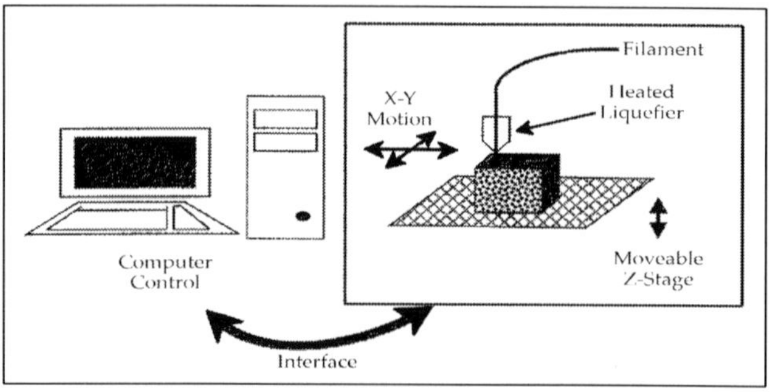

Figure 7.1 *I*llustration of Fused Deposition Modeling emeraldinsight.com

Modeling, prototyping and production can be done by using Fused Deposition Modeling an additive manufacturing technology. The basic principle of FDM is "additive" which means materials is laying down in layers. A plastic string or metallic wire from a coil is unwounded and supplies depositing material to an extrusion nozzle which can turn the flow on and off. The nozzle heats the material and able to move in both axes horizontal and vertical by a numerically controlled system, directly controlled by a computer –aided manufacturing (CAM) software. The model or part is made by extruding tiny drops of thermoplastic material to make a layer as the material solidify instantly after the extrusion from the nozzle. Movement of extrusion head is controlled by a special motors either stepper or servo.

FDM, a noticeable for fast prototyping, is used for prototyping and quick manufacturing. Quick prototyping facilitates iterative testing, and for very short runs, fast engineering can be a relatively cheap alternate. (David Ian Wimpenny, 2016)

Advantages: Inexpensive since uses plastic, more costly models use a different (water soluble) material to eradicate supports completely. Even low-cost 3D printers have enough resolution power for many applications.

Disadvantages: Supports leave marks that involve removing and sanding. Warping, narrow testing allowed due to Thermo plastic material.

SLA – Stereo Lithography

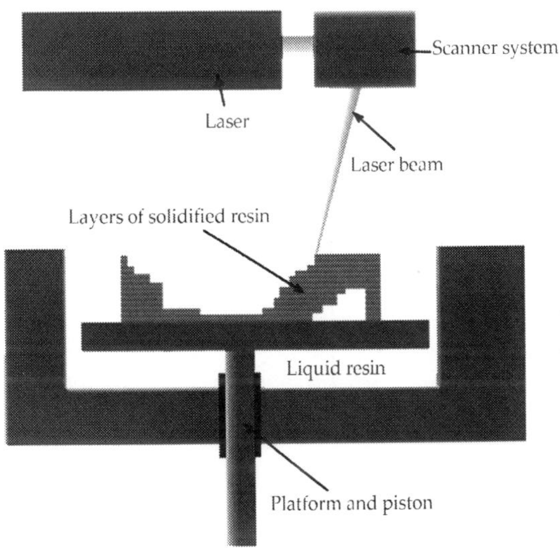

Figure 7.2 Stereo lithography *http://proto3000.com*

A process in which a vat of fluid ultraviolet cured photopolymer "resin " and an ultraviolet laser to form parts' layers each one at a time. This process is known as Stereo Lithography an additive manufacturing technique. In each layer, the laser beam followed a cross section of the part traced on the upper surface of fluid resin. On exposure of ultraviolet laser light, the pattern drawn on the resin cured and solidifies and join it to the layer beneath.

After the formation of pattern, the SLA's elevator platform inclined by a distance of one layer thickness, normally 0.05 to 0.15 mm (0.002"-0.006"). After that a blade full of resin sweeps on the surface of the part, re-coating it if required for fresh material. A successive layer pattern is drawn joined to the previous one. In this way a complete 3-D part is manufactured. On the completion of whole process the parts are dipped in a chemical bath for removal of excess resin and then cured in an ultraviolet oven. (Chee Kai Chua, 2014)

Supporting structures are required in Stereo lithography. These supports are used to attach the part to elevator platform, avoid deflection caused by gravity and fixe the cross section in the place so that they can support the horizontal pressure of coating blade. During the preparation of 3D Computer Aided models, the supports are generated automatically. They may be operated manually. Supports have to be removed by from the finished product physically, dissimilar to other cheap and fast prototyping methods.

Advantages and Disadvantages

Speed is main advantage of stereo lithography because useful components can be manufactured within a day. Time required to produce a typical component depends upon the size and complexity, it can be a few hours, few days or more. Stereo lithography can produce parts up to size of 50×50×60 cm (20"×20"×24"). Some SLAs like Mammoth stereo lithography machine have a platform 210×70×80 cm have the ability to produce part more than 2 m long. Stereo lithography can produce strong prototypes that be used as master pattern for injection molding, thermoforming, blow molding, and various metal casting processes. (David Ian Wimpenny, 2016)

Although stereo lithography can manufacture a wide diversity of shapes, it has often been costly; the cost of photo-curable resin has long ranged from $80 to $210 per liter, and the cost of stereo lithography machines has ranged from $100,000 to more than $500,000.

Cheaper SLA 3D printers have been created recently and one can only assume that in the future more will be created that are within the price range of individuals.

SLS - Selective Laser Sintering

An added substance producing system. A powerful laser (CO_2 laser) is utilized as a part of Selective laser sintering to dissolve small particles of plastic, metal (direct metal laser sintering), fired, or glass powders in a material that has required 3D shape. The laser just melts powdered material by scanning cross-areas delivered from a 3-D computerized outline of the part (for instance from a CAD record or output information) on the surface of a powder bed. The powder layer is dropped in the wake of scanning the cross area and another layer of material is framed on the top. This procedure is rehashed up to the fruition of the part.

A SLS machine as a rule utilizes a pulsed laser in light of the fact that completed part density relies upon peak laser power, as opposed to laser interval. The SLS machine warm up the bulk powder material in the powder bed as opposed to underneath its liquefying point, to make it is quiet for the laser to expand the temperature of the specific areas whatever remains of the route to the softening point.

A few SLS machines utilize single-part powder, for example, direct metal laser sintering. Be that as it may, most SLS machines utilize two-part powders, commonly either covered powder or a powder blend. In single-segment powders, the laser dissolves just the external surface of the particles (surface softening), intertwining the strong non-liquefied centers to each other and to the past layer. (Chee Kai Chua, 2014)

Contrasted and different strategies for added substance fabricating, SLS can create parts from a moderately extensive variety of industrially accessible powder materials. These incorporate polymers, for example, nylon (flawless, glass-filled, or with different fillers) or polystyrene, metals including steel, titanium, compound blends, and composites and green sand. The physical procedure can be full liquefying, halfway dissolving, or fluid stage sintering. Contingent upon the material, up to 100% density can be accomplished with material properties equivalent to those from customary assembling techniques. Much of the time substantial quantities of parts can be stuffed inside the powder bed, permitting high efficiency. (David Ian Wimpenny, 2016)

SLS is performed by machines called SLS frameworks. SLS innovation is in wide use the world over because of its capacity to effortlessly make exceptionally complex geometries directly from advanced CAD information. While it started as an approach to assemble model parts right on time in the plan cycle, it is progressively being utilized as a part of constrained run assembling to deliver end-utilize parts. One less expected and quickly developing utilization of SLS is its utilization in workmanship.

Benefits

SLS has many advantages over conventional production strategies. Speed is the most evident on the grounds that no uncommon tooling is required and parts can be worked in a matter of hours. Also, SLS takes into consideration more thorough testing of models. Since SLS can utilize most combinations, models would now be able to be useful equipment made out of an indistinguishable material from production parts.

SLS is additionally one of only a handful couple of added substance fabricating innovations being utilized as a part of production. Since the parts are built layer by layer, it is possible to plan hidden components and entries that couldn't be thrown or generally machined. Complex geometries and assemblies with numerous parts can be improved to fewer parts with a savvier get together. SLS does not require extraordinary tooling like castings, so it is helpful for short creation runs.

Applications

This technology is utilized to fabricate direct parts for a collection of projects including aviation, dental, medicinal and different businesses that have little to medium size, profoundly complex parts and the tooling industry to make direct tooling implants. With an assemble covering of 250 x 250 x 185 mm, and the capacity to "develop" numerous parts at one time, SLS is an exceptionally cost and time compelling technology. The technology is utilized both for quick prototyping, as it diminishes improvement time for new items, and generation producing as a cost sparing strategy to streamline congregations and complex geometries. (David Ian Wimpenny, 2016)

Constraints

The parts of size include points of interest and surface complete, and in addition print through mistake in the Z axis may be elements that ought to be considered preceding the utilization of the technology. Nonetheless, by arranging the work in the machine where most components are worked in the x and y axis as the material is set out, the element tolerances can be overseen well. Surfaces usually must be polished to accomplish reflect or extremely smooth completions.

For production tooling, material density of a completed part or embed ought to be tended to preceding use. For example, in injection molding embeds, any surface flaws will cause defects in the plastic part, and the supplements should mate with the base of the form with temperature and surfaces to avoid issues.

In this procedure metallic support structure expulsion and post handling of the part created is a time expending process and requires utilization of EDM and/or grinding machines having a similar level of accuracy given by the RP machine. (Chee Kai Chua, 2014)

PRESENT AND FUTURE APPLICATIONS OF 3D PRINTING

Biomedical Engineering

Recently researchers and specialists have already had the capacity to utilize 3D printing technology to make body parts and parts of organs. The main complete organ prepared through 3D Printing is probable to be done in the coming years. The way toward manufacturing the organ or body part is just the same as anyway you were to make a plastic or metal part, nevertheless, rather the unrefined material utilized are natural cells made in a lab. Scanning the patient's body and existing bone structure specifically for a particular patient, one can make certain that the patient's body won't dismiss the organ.

Another utilization of 3D printing in the biomedical field is that of creating limbs and other body parts out of metal or assorted materials to displace lost or harmed limbs. Prosthetic limbs are required in many parts of the world because of wounds managed amid war or by ailment. Presently prosthetic limbs are very costly and normally are not modified for the patient's requirements. 3D printing is being employed to sketch and express custom prosthetic limbs to meet the patient's exact requirements. By scanning the patient's body and present bone structure, inventors and architects can re-design the lost piece of that attachment. (Chee Kai Chua, 2014)

Aerospace and Automobile Manufacturing

High technology organizations, for example, aviation and vehicle makers have been utilizing 3D printing as a prototyping apparatus for quite a while. In any case, in recently years, with advance headway in 3D printing technology, they have possessed the capacity to make useful parts that can be utilized for testing. This procedure of design and 3D printing has enabled these organizations to propel their outlines speedier than at any other time because of the huge decreasing in the design cycle. From what used to take months amongst outline and the physical prototype, now inside hours the design group can have a prototype in their hands for checks and testing. (David Ian Wimpenny, 2016)

The eventual fate of 3D printing in these companies lies with making working parts directly from a 3D printer for use in the former item, not only to test purposes. This procedure is already underway for future autos and flying machine. The way in which 3D printing works (making a section layer by layer) enables the architect to make the part exactly the way it should be to fulfill the job needing to be done. Extremely complex geometry can be easily made utilizing a 3D printer, taking into consideration parts to be lighter, yet more grounded than their machined counterparts.

Construction and Architecture

Architects and city planners have been using 3D printers to create a model of the layout or shape of a building for many years. Now they are looking for ways of employing the 3D printing concept to create entire buildings. There are already prototype printer systems that use concrete and other more specialized materials to create a structure similar to a small house. The goal is the replace many cranes and even construction workers with these printing systems. They would work by using the 3D design model created on CAD software, to create a layer by layer pattern on the building just as a normal 3D printer works today. Most of the innovation in this area will have to come from the creation of the appropriate materials.

Product Prototyping

The creation of another item is always one of that involves many iterations of the same design. 3D Printing revolutionized the industry by allows designers to create and the next day see and touch their design. Never again did it take several meetings for everyone to agree on one design to create, and then wait months for the actual part to arrive. Nowadays a version of each idea is created and the next day, all are looked into together, in this way giving the ability to compare and contrast each one's features.

Plastic parts for example require molds and tooling to be created, these custom parts are expensive to create, therefore one must be certain the part designed meets the prerequisites. With 3D printing you can create a part that will look and feel exactly like the finished item. A few parts can also be tried just as the real injection formed part would.

Designing for 3D Printing

All the parts created using a 3D printer should be designed using some kind of CAD software. This type of production depends mostly on the quality of the CAD design and also the precision of the printer. There are many types of CAD software available, some are free others expect you to buy the software or have a subscription. (David Ian Wimpenny, 2016) Deciding what type of CAD software is beneficial for you will rely upon the prerequisites of what you are designing. However for beginners, that simply want to learn CAD and create basic shapes and features, any of the free CAD software packages will do.

When designing a part to be 3D printed the following points should be remembered:

- The part should be a strong, that is, not just a surface; it needs a real volume.
- Creating very small, or delicate features may not be printed properly, this depends greatly on the type of 3D printer that is going to be used.
- Parts with overhanging features will require supports to be printed properly. This ought to be taken into account since after the model should be cleaned by removing the supports.

This may not be an issue unless the part is very delicate, since it may break.

- Be beyond any doubt to calibrate the 3D printer before using it, it is essential to guarantee that the part adheres properly to the manufacture plate. On the off chance that it doesn't, sooner or later the part may come free and ruin the entire print work.
- Some thought ought to be given to the orientation of the part, since a few printers are more precise on the X and Y axes, then the Z axis.

Table 7.1 showing all available types of 3D Printers

Type	Technologies	Materials	
Extrusion	Fused deposition modeling (FDM)	Thermoplastics (e.g. PLA, ABS), eutectic metals, edible materials	
ranular	Direct metal laser sintering (DMLS)	Almost any metal alloy	
	Electron beam melting (EBM)	Titanium alloys	
	Selective heat sintering (SHS)	Thermoplastic powder	

	Selective laser sintering (SLS)	Thermoplastics, metal powders, ceramic powders	
	Powder bed and inkjet head 3d printing, Plaster-based 3D printing (PP)	Plaster	
Laminated	Laminated object manufacturing (LOM)	Paper, metal foil, plastic film	
Light polymerized	Stereo lithography (SLA)	photopolymer	
	Digital Light Processing (DLP)	liquid resin	
Type	Technologies	Materials	
Extrusion	Fused deposition modeling (FDM)	Thermoplastics (e.g. PLA, ABS), eutectic metals, edible materials	
ranular	Direct metal laser sintering (DMLS)	Almost any metal alloy	

	Electron beam melting (EBM)	Titanium alloys		
	Selective heat sintering (SHS)	Thermoplastic powder		
	Selective laser sintering (SLS)	Thermoplastics, metal powders, ceramic powders		
	Powder bed and inkjet head 3d printing, Plaster-based 3D printing (PP)	Plaster		
Laminated	Laminated object manufacturing (LOM)		Paper, metal foil, plastic film	
Light polymerized	Stereo lithography (SLA)		photopolymer	
	Digital Light Processing (DLP)	liquid resin		

THE MAIN COMPONENTS OF A 3D PRINTER:

A 3D printer incorporates an arrangement of parts that work all the while to deliver the desirable yield from the information computerized record, the fundamental parts of a 3D printer are detailed beneath:

Print Bed (Tray)

This is the level surface where the 3D models are layered during printing. The print bed might be surrounding or warmed Depending on the filament sorts utilized as a part of the printer. Warmed print beds are utilized to keep the printed area of the print warm amid the layering procedure to anticipate twisting.

Extruder

The extruder is the part that pushes out and bolsters the plastic filament (or some other filament) into the 'hot-end'. Extruders are regularly fused into the hot-end, be that as it may, in a few sorts, it can be remote, pushing the filament through a tube, called a Bowden cable, into the hot-end. In a few sorts, a double extruder is utilized, which gives the capacity to print two distinct materials in the meantime. This additional element brings about expanded cost, as it requires an additional extruder and hot end. Shown in Fig. 1

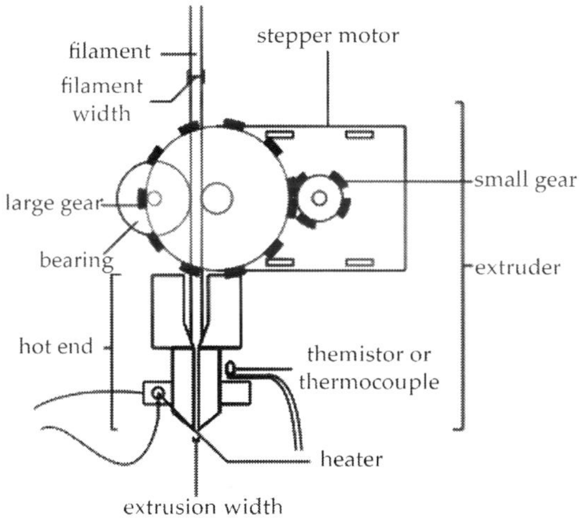

Figure 7.3 An extruder for 3 D printing 3DPrint.com

Hot-end

The hot-end is made out of a heat source, a temperature sensor, and an expulsion tip where the plastic filament is encouraged through to store liquid material, it is frequently mistaken for the extruder. The gap in the space may extend in the estimate, ordinarily in the vicinity of 0.2mm and 0.8mm. The littler the spout, the more point by point the print, yet the more it takes for the more slender layers to stack up.

Filament

The filament is the addition material which is figured as a 3D strong prototype by the printer. Like an inkjet imbue ink, a 3D printer transmits softened filament.

OPERATION OF A 3D PRINTER

The process of 3D printing (converting a digital file into compacted object) is a quite long and intricate one. This process is defined in the following 4 steps below:

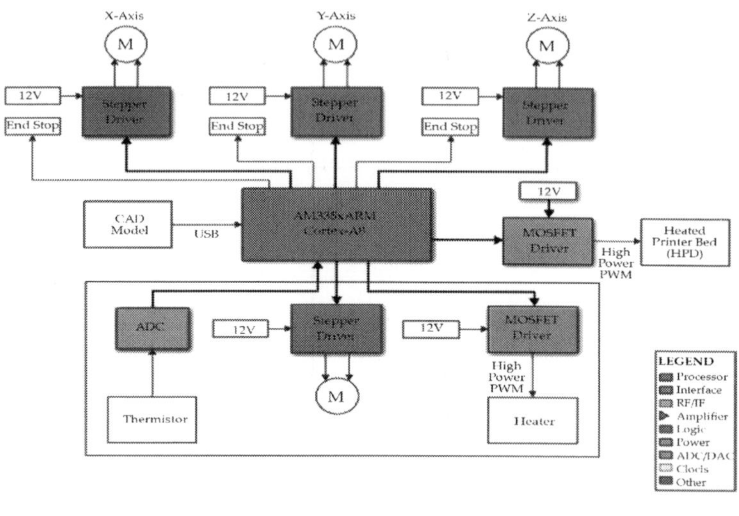

Figure 7.4 Block Diagram of 3D printer 3DPrint.com

Step I: (CAD)

(3) File — is prepared using a 3D modeling program, either without any

preparation or starting with a 3D display made by a 3D scanner. In any case, the program makes a document that is sent to the 3D printer. En route, programming cuts the plan into, at least hundreds likely thousands, of flat layers. These layers will be printed one on the other until the point that the 3D protest is finished. Many projects are employed for planning 3D protest, for example, Solid Works, AutoCAD and have a lofty expectation to absorb information, there are various different projects, many free that are anything but difficult to learn. The free form of Google Sketch Up, for instance, is exceptionally prevalent for its convenience; and the free Blender program is well known with its propelled highlights.

Step II: (motoring)

The working of 3D printer requires motoring with low torque, high accuracy the best machine to do this capacity is stepper motor: is an electromagnetic gadget that changes electronic beats into mechanical shaft revolution. Principal points of step motor are simplicity, high reliability, high torque at low speeds and a basic they are an exceptional type of synchronous motor which are intended to pivot a particular number of degrees for every electric pulse gotten by its control unit, a small amount of degree should be possible utilizing gears keeping in mind the end goal to move the printer head a particular relocation, in 3D printer four stepper motor were expected to do the particular capacity, three of them were utilized for moving in X,Y, Z direction of the printer head, the fourth one was expected to move the plate(bed). (David Ian Wimpenny, 2016)

Step III: (processing)

The AM335x chip are improved with picture, graphics preparing, and mechanical interface alternatives. The apparatus supports the associated high-level operating systems (HLOSs), The AM335x chip contains these subsystems (1) Microprocessor unit (MPU) (2) Graphics Accelerator subsystem for 3D graphics quickening to help display. Permitting free operation and timing for more noteworthy efficiency and flexibility. As the ability to scale in speed from 600MHz to more prominent than 1GHz. (Chee Kai Chua, 2014)

Step IV: (MOSFET drive)

Bipolar stepper engine with MOSFET drive, which will enables the engine to move in both direction, since stepper engine is open circle system, high accuracy part will utilized as a part of request to print in the particular place, so that MOSFET was utilized, on the grounds that it has a quick frequency operation.

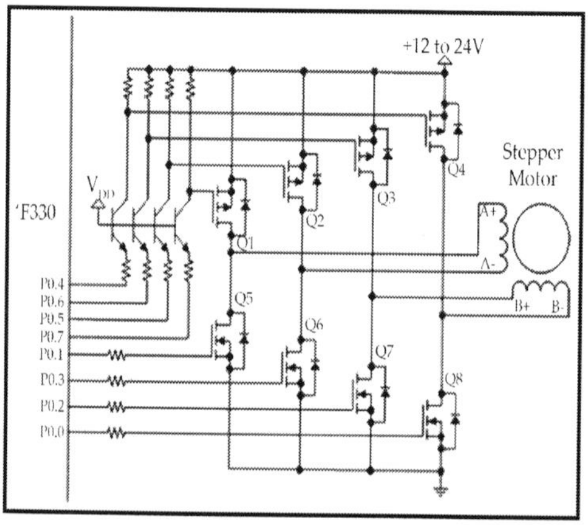

Figure 7.5 MOSFET Drive of stepper motor. 3DPrint.com

DIFFERENT TYPES OF 3D PRINTERS

FFF/FDM 3D printers take many carry out. They vary in mechanical devices and facilitate structures.

A couple of makers pick mechanical simplicity to the inconvenience of moving to create & organize while others endeavor to increase extruder head speed by using settled motors and complex belt initiation. The most common mechanical strategies for FFF/FDM 3D printers are:

- Cartesian-XY-head
- Cartesian-XZ-head
- Delta
- Core XY
- Polar
- Scara (robot arm)

Cartesian-XY-Head

The first RepRap 3D printer, the Darwin, was based on the Cartesian-XY-head organization. The extruder head travels over the X and Y axis and the bed above the Z.

Z axis movement on such a 3D printer is very precise and requires very low accelerations, but the bed needs to be lightweight in order to maintain accuracy, which makes it more difficult to add a fully automatic bed leveling system.

The best specimen of a Cartesian-XY-head 3D printer is the: Stacker S2.

Cartesian-XZ-Head

The Cartesian-XZ-head planning was first lead by the Mendel which was another version of the original RepRap – the Darwin. This procedure varies form artesian-XY-head because it moves the bed over the Y axis and the extruder head over the X axis and the Z axis.

The main advantage of this arrangement is that the bed can grasp a lot of weight, making it conceivable to add a (heavy) fully automatic bed leveling system (see video below).

Examples of Cartesian-XZ-head 3D printers are: Original Prusa i3 MK2 and RepRap Guru Prusa i3 V2.

Delta

Delta 3D printers also work within the Cartesian plane, though the system of the frame is totally dissimilar. They are called Delta because the extruder head is on hold by three arms in a triangular formation. Besides that they have a circular print bed. The advantage of a Delta 3D printer is that the poignant parts are trivial and therefor limit the inertia. That results in quicker printing with greater accuracy.

An example of a Delta print is the Rostok Max v3.

Core-XY

A sample of a Core-XY printer is: Airwolf 3D AW3D AXIOM.

Core-XY is a Cartesian procedure that is quickly developing in popularity. The drive on the XY gantry depends on a combined effect of X and Y motors. Core-XY is a similar manipulator system, which means that the motors on a Core-XY system are fixed. Corresponding manipulator systems give more quick acceleration than successive stack-up arrangements like Cartesian-XZ-head. The video below shows you how blazing fast a Core-XY setup can be.

Polar

Polar 3D printers have a spinning print bed, plus an extruder head that can travel left, right up and down.

A polar 3D printer is energy effective because it only needs two stepper motors in conflicting to for instance a Cartesian planning which requires a minimum of one stepper motor for each axis, so commonly minimum four. Principally, it is really cool to see in action.

An example of a Polar printer is: Polar 3D.

3D PRINTING MATERIALS

- The materials accessible for 3D printing have made some amazing progress since the beginning of the innovation. There is currently a wide assortment of various material sorts that are provided in various states (powder, fiber, pellets, granules, gum and so forth).
- Specific materials are currently for the most part created for particular stages performing committed applications (a case would be the dental area) with material properties that all the more correctly suit the application.
- However, there are currently far an excessive number of exclusive materials from the various 3D printer merchants to cover them all here. Rather, this article will take a gander and no more prominent sorts of material in a more nonspecific way. And furthermore two or three materials that emerge.

Plastics

Nylon, or Polyamide, is generally utilized as a part of powder frame with the sintering procedure or in fiber shape with the FDM procedure.

It is a solid, flexible and durable plastic material that has demonstrated dependable for 3D printing. It is obviously white in shading yet it can be hued — pre-or post-printing. This material can likewise be consolidated (in powder arrange) with powdered aluminum to create another normal 3D printing material for sintering.

- ABS is another basic plastic utilized for 3D printing, and is broadly utilized on the section level FDM 3D printers in fiber shape. It is an especially solid plastic and arrives in an extensive variety of hues. ABS can be purchased in fiber shape from various nonproprietary sources, which is another motivation behind why it is so well known. (Chee Kai Chua, 2014)
- PLA is a bio-degradable plastic material that has picked up footing with 3D printing for this very reason. It can be used in gum design for DLP/SL forms and in addition in fiber shape for the FDM procedure. It is offered in an assortment of hues, including straightforward, which has ended up being a valuable alternative for about 3D Printing Materials: Plastics a few uses of 3D printing. Be that as it may it is not as solid or as adaptable as ABS.
- LayWood is an exceptionally created 3D printing material for section level expulsion 3D printers. It comes in fiber frame and is a wood/polymer composite (additionally alluded to as WPC). 3D Printing Materials: Metals A developing number of metals and metal composites are utilized for modern review 3D printing. Two of the most widely recognized are aluminum and cobalt subsidiaries.
- One of the most grounded and in this manner most usually utilized metals for 3D printing is Stainless Steel in powder frame for the sintering/dissolving/EBM forms. It is actually silver, yet can be plated with different materials to give a gold or bronze impact.
- In the most recent few years Gold and Silver have been added to the scope of metal materials that can be 3D printed straightforwardly, with clear applications over the adornments area. These are both exceptionally solid materials and are handled in powder shape.

- Titanium is one of the most grounded conceivable metal materials and has been utilized for 3D printing modern applications for quite a while. Provided in powder shape, it can be utilized for the sintering/liquefying/EBM forms.

Earthenware Production

Earthenware production are a generally new gathering of materials that can be utilized for 3D printing with different levels of accomplishment. The specific thing to note with these materials is that, post printing, the fired parts need to experience an indistinguishable procedures from any fired part made utilizing customary techniques for creation — in particular terminating and coating.

Paper

Standard A4 copier paper is a 3D printing material utilized by the exclusive SDL prepare provided by Mcor Technologies. The organization works a strikingly extraordinary plan of action to other 3D printing sellers, whereby the capital expense for the machine is in the mid-extend, yet the accentuation is particularly on an effortlessly realistic, financially savvy material supply, that can be purchased locally. 3D printed models made with paper are protected, earth cordial, and effectively recyclable and require no post-preparing.

Bio Materials

There is a colossal measure of research being directed into the capability of 3D printing bio materials for a large group of medicinal (and other) applications. Living tissue is being explored at various driving organizations with a view to creating applications that incorporate printing human organs for transplant, and also outside tissues for substitution body parts.

Other research around there is centered around creating nourishment stuffs — meat being the prime case.

Nourishment

Trials with extruders for 3D printing sustenance substances have expanded significantly finished the most recent few years. Chocolate is the most widely recognized (and alluring). There are additionally printers that work with sugar and a few examinations with pasta and meat. Looking to the future, inquire about is being attempted, to use 3D printing innovation to create finely adjusted entire dinners.

Other

Lastly, one organization that has a one of a kind (restrictive) material offering is Stratasys, with its computerized materials for the Objet Connex 3D printing stage. This offering implies that standard Objet 3D printing materials can be joined amid the printing procedure — in different and indicated focuses — to frame new materials with the required properties. Up to 140 diverse Digital Materials can be acknowledged from consolidating the current essential materials in various ways.

GLOBAL EFFECTS ON MANUFACTURING

3D printing is as of now affecting how items are fabricated – the nature of the innovation allows better approaches for speculation as far as the social, monetary, ecological and security ramifications of the assembling procedure with all around good outcomes.

- One of the key elements behind this announcement is that 3D printing can possibly convey generation nearer to the end client as well as the buyer, subsequently lessening the present production network limitations. The customisation estimation of 3D printing and the capacity to create little generation clusters on request is a certain approach to connect with purchasers AND decrease or nullify inventories and stock heaping — something like how Amazon works its business. (Chee Kai Chua, 2014)

Transportation save parts and items from one a player on the planet to the next could conceivably wind up noticeably out of date, as the extra parts may perhaps be 3D imprinted nearby.

This could majorly affect how organizations extensive and little, the military and buyers work and communicate on a worldwide scale later on. A definitive go for some is for shoppers to work their own particular 3D printer at home, or inside their group, whereby advanced outlines of any (adaptable) item are accessible for download by means of the web, and can be sent to the printer, which is stacked with the right material(s). Right now, there is some level headed discussion about whether this will at any point happened, and considerably more thorough civil argument about the time period in which it might happen.

- ✓ The more extensive appropriation of 3D printing would likely cause re development of various as of now imagined items, and, obviously, a much greater number of totally new items. Today beforehand inconceivable shapes and geometries can be made with a 3D printer, however the trip has truly just barely started. 3D printing is accepted by many to have exceptionally incredible potential to infuse development into advancement and bring back neighborhood fabricating.

  The utilization of 3D printing innovation effectsly affects the worldwide economy, if received around the world. The move of creation and dispersion from the present model to a restricted generation in light of interest, nearby, tweaked generation model could possibly decrease the awkwardness amongst fare and import nations.

- ✓ 3D printing would can possibly make new enterprises and totally new callings, for example, those identified with the generation of 3D printers. There is an open door for proficient administrations around 3D printing going from new types of item architects, printer administrators, material providers the distance to licensed innovation lawful question and settlements. Theft is a present concern identified with 3D printing for some IP holders.

- ✓ The impact of 3D imprinting on the creating scene is a double-edged sword. One case of the constructive outcome is brought down assembling taken a toll through reused and other neighborhood materials, yet the loss of assembling occupations could hit many creating nations extremely, which would set aside opportunity to overcome. (Chee Kai Chua, 2014)

- ✓ The created world, would profit maybe the most from 3D printing, where the developing matured society and move of age socioeconomics has been a worry identified with generation and work constrain. Likewise the medical advantages of the therapeutic utilization of 3D printing would provide food well for a maturing western culture. Effect on Commercial Manufacturing

While it is hard to state with assurance how 3D imprinting in its different structures (e.g., desktop, extensive scale) will affect conventional assembling, rising patterns show that a major outlook change has as of now began. As 3D printing develops, the new universe of assembling resembles this: • Time-to-advertise for items shrivels. This will be expected, to a limited extent, to speedier outline and prototyping cycles because of 3D printing, additionally to the disposal of tooling and industrial facility setup times for new items. Being "lithe" will never again be an upper hand yet a fundamental need to remain in business.

- ✓ Products have prevalent capacities. The hindrances for assembling will be brought down, bringing new contenders with new thoughts. In the meantime, items joining 3D-printed parts will show prevalent components, for example, being littler, lighter, more grounded, less mechanically intricate and less demanding to keep up. These items will hold particular upper hand. • Open plan is digging in for the long haul. Groups of end clients will progressively be in charge of item plans, which will be accessible to anybody with the essential abilities and apparatuses who needs to outline and after that produce. These open-outline items will be better than restrictive items. Makers will contend on how well they actualize the plans and their assemble quality, which will be barbarously evaluated by end clients on the Internet. • Customization is the new ordinary. As imaginative organizations utilize 3D printing and other fast procedures to offer customization at no extra cost, buyers will generally expect customization as the standard. The per-unit fabricating expenses of little generation runs (even clumps of one) will approach those of long keeps running as innovation obstructions fall. • The financial matters of seaward change. The value advantage related with large scale manufacturing in ease locales will be tested by 3D printers giving in the nick of time fabricating close to the point of offer or purpose of get together.

> Supply chains will be re-advanced to consider the benefits of in the nick of time, especially for low-volume or exceedingly particular segments. Then again, fashioners will have the capacity to limit costs by utilizing minimal effort, high-volume segments wherever conceivable, associated with particular Justin-time segments delivered at the purpose of get together. In the midst of this new universe of assembling, conventional assembling forms must develop or bite the dust. (See sidebar.) In a current report, LEF analyst Simon Wardley noticed that when an action, for this situation fabricating, turns into an item, conventional practices must advance to grasp the new, however profoundly problematic, business forms. He expresses that the 3D printing disturbance "will probably be driven by new participants whose practices will be profoundly unique in relation to those of existing players."70 Therefore, in get ready for this change, customary producers must stay up to date with advancing 3D printing hones and know about their own particular inner boundaries (e.g., culture, association) that could keep them from exploiting the change. (Chee Kai Chua, 2014)

As more associations and people progress toward becoming producers, the lines amongst maker and client will obscure. At the point when there is a retailer in the middle of, those lines will obscure as well. Assembling will move into retailing. Buyers and new participants will move into assembling. Will customary assembling be dead in 10 years? No, however it will look altogether different.

CHAPTER 8

Manufacturing Process Management

INTRODUCTION

Process management is the utilization of learning, abilities, instruments, strategies and frameworks to characterize, envision, measure, control, report and enhance forms with the objective to meet client necessities productively. Assembling process administration (MPM) is an accumulation of advances and strategies used to characterize how items are to be made. Assembling process administration characterizes "how" an item is to be created MPM gets the advanced item definition from designing, considers the abilities and limits of inside plants and outside providers, and conveys the arrangement of assembling designs required to deliver the item. MPM likewise involves giving early criticism from assembling to building in light of preparatory forms of the assembling forms. Also, the MPM procedure speaks with generation frameworks (ERP/MES), conveying advanced routings, and also fabricating bills-of-materials (mBOMs) and work directions, including every supporting report required by creation administrators to assemble the item. (Alistair Barros, n.d.)

UNDERSTANDING THE NEED TO OPTIMIZE MANUFACTURING PROCESS MANAGEMENT

Upgrading the MPM procedure is winding up noticeably progressively vital to organizations. Sadly, the procedure itself is additionally winding up plainly more hard streamlining. Items, while winding up noticeably more mind boggling, must stay configurable to address advertise issues. In the meantime, generation volume, cost, and time-to-showcase targets are contracting. Assembling associations must adjust their assembling condition to increase new outlines with higher profitability, flexibility and quality. There are four key territories where organizations can commonly enhance their MPM procedure:

1. Define the Manufacturing Process During the Product Design Phase. For the most part, planners "toss the outline over the divider" to assembling engineers. Be that as it may, if fabricating engineers were given direct access to the steadily developing outline data, they could define the assembling procedure in parallel with item plan.

   Moreover, with this data, assembling will turn out to be better prepared to deal with simultaneous item advancement and better ready to adjust to in-handle designing changes.

2. Provide Manufacturing engineers with access to designing Data. Precise assembling process definition requires that assembling engineers have the capacity to specifically reuse designing information, including parts, classification, 3D mockups, and producing necessities, for example, GD&T data. However, commonly, producing engineers need guide access to this data. (Alistair Barros, n.d.)

3. Dispense with Manual Processes. Today, most process designs are defend utilizing a spreadsheet program, and work directions are made utilizing a word processor. This utilization of manual apparatuses prompts an assortment of issues: moderate execution of the procedure, information duplication and mistake, and disarray at whatever point a change strikes the designing definition of the item.

4. Enhance undertaking coordinated effort. Since assembling situations regularly traverse numerous plants and time zones, and with makers looking to execute "outline anyplace, construct anyplace" procedures, it's difficult for assembling architects to catch and offer their insight, and guarantee consistency over the venture.

FOCAL POINTS OF STREAMLINING MANUFACTURING PROCESS MANAGEMENT

Typical focal points from improving the MPM method include:

- ✓ Improved capability of amassing engineers
- ✓ Enables organize reuse and control of building information in get ready designs, to avoid data duplication
- ✓ Enables reuse of regulated and institutionalized techniques and resources
- ✓ Reduces time required to make and revive work bearings however their dynamic period

IMPROVED PRODUCTION INCREMENT AND PRODUCTIVITY

- ✓ Reduces get ready time and desire to learn and adjust with correct and imperative 3D work rules
- ✓ Reduces creation experimentation strategy for streamlining delivering frames with cutting edge endorsement lessened cost of changes
- ✓ Identifies required layout changes before in the arrangement, with advantageous contribution from gathering
- ✓ Streamlines change influence unmistakable verification and spread
- ✓ Increases building detectable quality to the potential gathering

impact of a change, to energize fiscally wise arrangement decisions

- ✓ Provides a key change organization structure, which reinforces both outlining and gathering needs diminished piece and modify
- ✓ Reduces the threat of making mixed up thing courses of action by wiping out irregularities between the latest procedure definition and the work headings used on the shop floor

THE MPM PROCEDURE STEPS

1. Characterize Manufacturing system. In this progression, fabricating engineers assess both the plan prerequisites and the assembling abilities and limits important to help the assembling procedure. They distinguish which accomplices will be required, and any long lead things that will require prompt consideration.

2. Propelled Manufacturing Planning. ("Progressed", in this unique situation, signifies "ahead of schedule") In this progression, fabricating engineers work simultaneously with building, as right on time as conceivable in the plan procedure, utilizing preparatory outline data. Together, they make preparatory renditions of the assembling bill-of-material (mBOM) and process designs, start new tooling solicitations, and create time and cost gauges. Accurate input or demands for-change are likewise given to building in the early phases of item configuration to enhance manufacturability.

3. Upgrade Process Plan. In this progression, producing engineers make a few assembling process contrasting options to look at and upgrade the assembling procedure. Assembling process choices can speak to either unique options in one plant, choices between various plants, or make-versus-purchase alternatives. Streamlining strategies commonly incorporate line adjusting, yet can likewise include more specific sorts of recreation, for example, work-focus reproduction or entire plant discrete occasion reenactment.

4. Archive Process Plan. Once the procedure design and item con-

figuration are develop, the assembling engineer must detail and record the assembling forms that have been chosen from the past enhancement step. PTC's Pro/ENGINEER® or other 3D CAD/CAM instruments are utilized to define NC (numerical control) device ways or CMM (organize measuring machine) assessment programs. Work guidelines are delivered for the administrator, and normally incorporate pictures of the item, which are powerfully created from the procedure definition with Windchill MPMLink. These pictures can be cooperatively defined utilizing Pro/ENGINEER or PTC's ProductView® representation arrangement, with the goal that they may effectively be stayed up with the latest. (Alistair Barros, n.d.)

5. Pre-creation and Production. In this progression, the assembling procedure definition is electronically discharged to generation frameworks (ERP/MES). Work directions are made accessible to the shop floor in either printed or electronic arrangement. Electronic work guidelines can likewise be utilized to gather criticism from the genuine execution of the procedure, for example, deviations, waivers and as-fabricated BOMs. (Alistair Barros, n.d.)

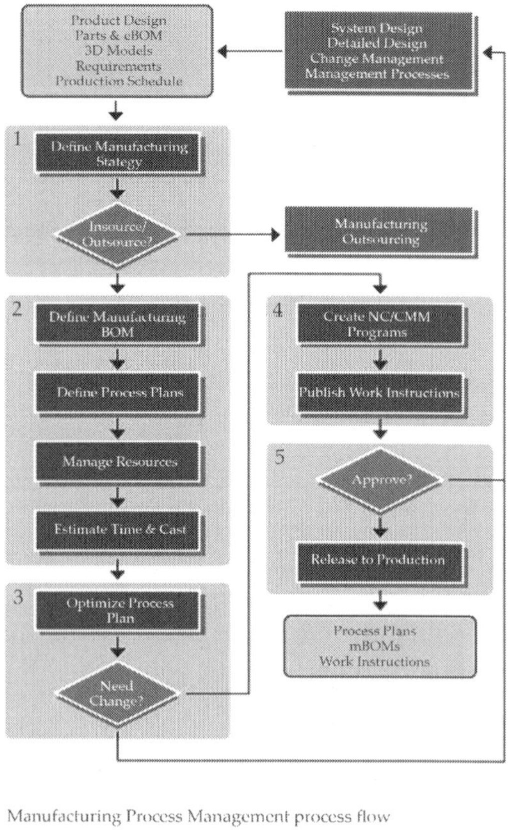

Figure 8.1 Manufacturing Process Management Flow chart (Bessant, n.d.)

CELLULAR MANUFACTURING

Cellular manufacturing is rising as one of the significant devices being utilized as a part of the drive to assembling intensity. Cellular Manufacturing is a procedure of assembling which is a subsection of without a moment to spare assembling and lean assembling including bunch innovation. The objective of Cellular Manufacturing is to move as fast as could reasonably be expected, make a wide assortment of comparable items, while making as meager waste as could be allowed. Cell producing includes the utilization of numerous "cells" in a mechanical production system form. Each of these cells is made out of one or numerous distinctive machines which achieve a specific assignment. The item moves starting with one cell then onto the next, each station finishing some portion of the assembling procedure. Frequently the cells are masterminded in a "U-shape" outline since this takes into account the supervisor to move less and can all the more promptly watch over the whole procedure. One of the greatest points of interest of cell producing is the measure of adaptability that it has. Since the greater part of the machines are programmed, straightforward changes can be made quickly. This takes into consideration an assortment of scaling for an item, minor changes to the general plan, and in extraordinary cases, completely changing the general outline. These progressions, albeit dull, can be proficient to a great degree rapidly and precisely.

A cell is made by solidifying the procedures required to make a particular yield, for example, a section or an arrangement of guidelines. These cells take into account the lessening of superfluous strides during the time spent making the particular yield, and encourage brisk distinguishing proof of issues and support correspondence of representatives inside the phone to determine issues that emerge rapidly. Once actualized, Cellular Manufacturing has been said to dependably make huge picks up in profitability and quality while at the same time decreasing the measure of stock, space and lead time required to make an item. It is hence that the one-piece-stream cell has been called "a definitive in lean creation

The notoriety of cell make mirrors the flighty idea of current business. Adaptability, fast reaction and opportunity to modify bearing are survival issues. As per Ingersoll Engineers (1993), 73% of UK organizations utilize Cellular Manufacturing. It enables organizations to modularize their exercises what's more, decay energy to bring down levels in the association. This application is some of the time depicted as a 'business-inside business' approach. Many organizations are following the pattern towards item focused cells, with the hardware, multi-talented individuals and bolster capacities expected to create a completed item, sub-get together or part.

The current overview by Goodyer and Higgins (1993) demonstrated that organizations' primary destinations in presenting cell producing are:

- On time conveyance
- Improved client reaction
- Reduced stock

- Improved quality

These are the "hard" business drivers important to mechanical creation. They relate fundamentally to enhanced market reaction and better coordination and designing. Quality change is one of the fundamental goals for presenting cell produce, and one of the key issues to handle while increasing aggressive advantage.

Cellular Manufacturing Results

The Ingersoll survey (1990), research, Wemmerlov and Hyer (1989), and different logical examinations, Lloyd and Mason (1993) and Caulkin (1988) have created the impression that associations are meeting their basic goals, and that imperative updates have been made, particularly, in movement times and diminished inventories. Consistently these are:

- Reduced inventories by 30%
- Reduced lead times significantly

Notwithstanding, in spite of the way that the impressive triumphs lie in the operational areas of associations (as appeared) liberal quality upgrades are generally avoiding various cell foundations. Again and again world centered levels of significant worth have been expert by coming about anguishing changes to the arrangement of the cell and the way it is worked.

Despite these "troublesome" issues, cell creating is similarly having a broad influence in people issues. Generally, cells are making another kind of agent, one who has distinctive aptitudes, whose endeavors may include: machine working, quality control, booking, bolster, the once-over is never-endingly growing. The affirmation is that the effect for the people who work in the cells is that their occupations are directly all the more asking for, captivating and satisfying.

The Ingersoll (1990) and Goodyer and Higgins (1993) thinks about have highlighted the essentialness of arranging people to understand and recognize the new strategy for working and to have the fundamental aptitudes. Typical answers highlights the essentialness people have on the achievement of a cell:

- Cell delivering is as much about people as machines"
- On those areas where the cell approach is beginning to work it

is required to the shop floor worker"

- Workers must be allowed commitment in cell and process design or they won't guarantee the cells"

It can be presumed that generally the "hard" business objectives are the drivers for cell create; in any case it is the "fragile" people issues which will ensure accomplishment.

Cellular Manufacture—the Shortfalls and their Remedy

As already discussed cell manufacturing is to an awesome degree standard and is getting benefits in various business regions, however its insufficiency is quality. Beginning investigation has highlighted that the clarifications behind this quality difficulty are that the arrangement technique of the cell is insufficiently revolved around portraying and achieving specific business benefits.

Numerous devices for the framework and organization of things and techniques for high caliber have been introduced throughout the latest couple of years. However there has been no fruitful general arrangement theory for use in a 'Simultaneous Engineering' mode to diagram cells and their things. (Alistair Barros, n.d.)

The standard issues which associations have seen to be troublesome are:

- Breakdown of whole thing and whole generation line particulars to points of interest for parts and cells.
- Design of occupations to help people to make earnestly awesome quality.
- Design of process controls for high combination short bundle create.
- Establish 'beat point' estimation and specifying frameworks for quality estimation in a system.

Starting at now work has been grasped at Coventry University to make a procedure whereby thing, process and occupations are arranged at the same time, insinuate Figure 1. This will enable customer needs and ought to be the driver all through the cell configuration handle, bringing about world aggressive quality right on time in the lives of the phones.

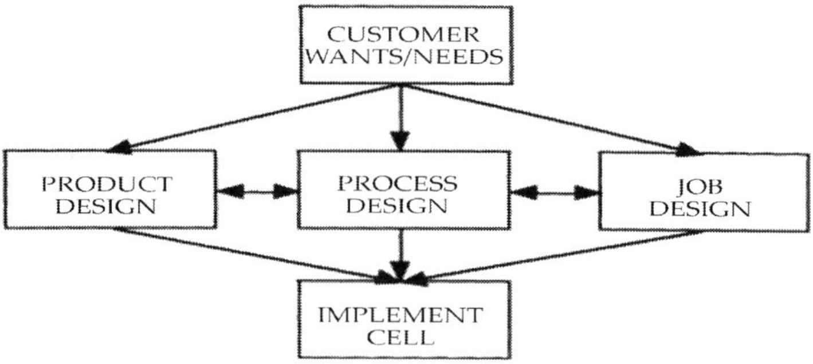

Figure 8.2*:* Concurrent Cell Design (Bessant, n.d.)

Work Design

As effectively indicated, people have the most basic effect on the accomplishment or, of course frustration of a cell. Generally, various foundations have point by point various unimaginable results of people working in cells. For example cells can make:

- Ownership
- Autonomy
- Flexibility

These results are frequently 'all in or hard and fast'. A couple of foundations outline the aptitudes that are required, using capacity structures and setting up the inclination deficiencies, others give get ready workshops on basic deduction systems and gathering building. Nonetheless, various foundations don't have a composed approach while plotting jobs, they are by and large progressions of past occupations, slanting towards free-play. The issues which can develop are:

There is a "dull" zone about social orders' obligations (cell limits/ support limits are not all around portrayed). Ordinarily cell specialists' obligations are left especially open: they ought to be as versatile as would be judicious.

- Usually there is no learning limit joined with occupations.
- Generally existing occupation plots are associated, with little change, to the cell.
- There is no system for evaluating the sufficiency of the business

plot.

- Job designs are regularly created after the cell organize is arranged.

With a particular true objective to address these issues as an indispensable bit of the system, the goals must include:

'To design the total human undertaking and the individual errands required in a cell to fulfill its quality objectives'.

At each period of the rationality the occupation design process will describe what is required to achieve the general target. The request which ought to be tended to are:

- Estimation: What quality measures imply the occupation execution and extent of capacities required?
- How would we orchestrate work traces?
- Particular: Which centered quality issues ought to be portrayed to decide work framework and business content?
- Idea: What highlight of the occupations ought to be portrayed to ensure quality targets will be expert?
- How would we perceive capable and insufficient game plans?
- Usage: What steps are relied upon to execute the new occupation designs besides, ensure that they are achieving comes to fruition

CONCURRENT ENGINEERING

Concurrent building, generally called synchronous outlining, is a strategy for arranging and making things, in which the different stages run at the same time, rather than ceaselessly. It lessens thing headway time and moreover a chance to promote, inciting upgraded effectiveness and diminished costs. (Prasad, n.d.)

Synchronous Engineering is a whole deal business procedure, with whole deal points of interest to business. Regardless of the way that basic execution can be trying, the high ground suggests it is useful in the whole deal. It clears the need distinctive blueprint patches up, by making an area for arranging a thing right the primary go through round. Starting late, Western gathering associations have seen that they have to rethink their approach to manage collecting. This has risen as a result of stiffer all inclusive competition, particularly from the Far East.

In the East, associations can convey absolutely new consequences of a high gauge fundamentally speedier than has been for the most part possible by their Western accomplices. Such an ability to show things to the market quickly is an essential forceful favorable position, as studies have exhibited that the underlying two firms to enter a market can secure up to 80% of the total bit of the general business (Brazier and Leonard, 1990). It is, thusly, incredibly appealing to twist up detectably one of these market pioneers.

The arrangement methodology has been recognized as being of genuine importance to the power of a thing. It has been surveyed that in the region of 70 and 80% of the costs over the life of a thing are achieved at the change stage, and it is at this stage, subsequently, that critical points of interest can be recognized by diminishing change time to as mind blowing a degree as is possible without impacting thing execution.

Traditional approaches to manage the arrangement method have been progressive in nature. Each office inside an attempt "continues", so to speak, at dealing with the layout and after that passes it on to the accompanying stage, having accomplished what is believed to be a perfect solution as far as it matters for them of the thing. (Prasad, n.d.)

An average issue with this kind of approach is that the social affairs careful for setup are absolutely ignorant of the necessities of the amassing limits. Blueprints are made with by zero idea of how they will be delivered, and are then slighted to the collecting side of the business. Perhaps a few the diagrams conveyed along these lines proceed with straight through the handle. Exactly when stood up to with issues, the all-inclusive community in the gathering limits (who, by shot, are generally as clueless of the fashioners necessities as the fashioners are of theirs) either make minor acclimations to the arrangement themselves, or pass it back for revision. (Alistair Barros, n.d.)

Cycle, for instance, this is for all intents and purposes unavoidable in the midst of the change of new things in the standard, progressive path as not a lot of fashioners have the fitness and data to convey a blueprint which will be absolutely acceptable. As said some time recently, a critical degree of the cost of another thing can be credited to the progression stage, and this cycle increases the cost senselessly. One of the contributing factors to this may be that while each office has its own specific point of view of what a perfect course of action should be, the distinctive workplaces' answers don't generally contrast with the requirements of the others. Such a nonappearance of accord customarily prompts perplexity and a conflict of interests which can set aside some chance to decide. We should note, in any case, that disregarding its verifiable shortcomings, the traditional layout handle still achieves its focuses, yet that extended forceful weight has compelled us to find ways to deal with lessen waste at all levels.

Concurrent (or Simultaneous) Engineering is a thought which has seen restored interest as of late. A similar number of people have pointed out, Concurrent Designing (C.E.) is for the most part a progressive issue, yet it makes the feeling that the happening to new gadgets, (for instance, Q.F.D., and PC applications to support information dealing with) and centered weights have restored excitement for it. Distinctive definitions have been offered for C.E., yet a champion among the most by and large recognized is that of the U.S. Agency of Defense. They express that C.E. is "… the think procedure for all the while arranging both the thing and its downstream era and support processes."(Carter and Baker, 1991). This is proficient by the gathering of an approach to manage thing headway which incorporates bunches whose people address every limit which has a promise to make to the thing. Writing in this field customarily fuses limits, for instance,

Advancing, Engineering Design, Manufacturing Engineering, Engineering Investigation, Production Planning, and so forth in the thing headway handle. In a wander at exhibit being endeavored at Napier University, we are inspecting courses in which we can reinforce the practical sharing of information inside multi-disciplinary social events working at the same time. Remembering the true objective to set up the levels of correspondence which must be supported, we have expected to recognize a sensible methodology for showing the layout strategy which will give us and valuation for which parties are incorporated and when.

Modeling the Design Process

Many individuals, for example, (Pugh, 1991), Roth, Koller (Pahl and Beitz, 1988) and (Page, 1992), have proposed models of the plan procedure or item life cycle. In the fundamental, these models have been worried about the stages included in the advancement of an item as opposed to the elements of a business which are included. A few models are then stretched out by mulling over the means which are included at each stage. This can be found in Roth's model which we have picked for instance (see figure 3). Many models have been produced, however, the agreement among the different creators has all the earmarks of being that plans advance through a progression of stages in a straight and consistent way. (Prasad, n.d.)

Toward the start of the procedure, we see a need recognized, trailed by the era of a particular which will satisfy that need. Once the detail for the new item is built up, work begins on the making of outline ideas which are then assessed and chosen to advance to point by point plan. At the point when the outline is finished, make of the new item can be arranged, also, embraced. (Alistair Barros, n.d.)

Having inspected these models, we recommend that they can be utilized as a premise on which to construct new models for the simultaneous approach. The limits of the stages speak to points of reference by which the majority of the data required for the following stage to be finished ought to be accessible. For instance, it is broadly acknowledged that details should be characterized before outlines or ideas are created (it is conceivable to make ideas before a detail is resolved, however this leads to emphasis and squandered assets). On the off chance that a halfway particular exists, some plan work can begin, however the reality remains that some type of detail ought to exist in advance. A comparable arrangement of needs exists all through the item life cycle (the outline procedure is an intricate arrangement of between useful conditions).

The primary purpose of C.E. is the co-operation between disciplines with a view to getting the item "right first time" by having the required learning and skill accessible to settle on educated choices and bargains, subsequently lessening the requirement for cycle. It doesn't request that the succession of stages through which the item advances ought to in a general sense change, despite the fact that the boundaries between them may become less clearly defined (as illustrated in other models of C.E. where the phases appear to merge into one development phase incorporating all aspects of the life-cycle). To allow the model to illustrate Concurrent Engineering it has to be modified to suit our purposes. Instead of considering the process simply in terms of phases, we must extend the model to consider which functions are involved at each stage. Look at figure 4. This model is based on the tasks shown in Roth's and the findings of an initial study at a local company. At present, this "functional" model is based on assumptions of which functions become involved at each operation. (Alistair Barros, n.d.) A detailed study of the company is currently being undertaken in an effort to arrive at a sound model with relevance to an actual case

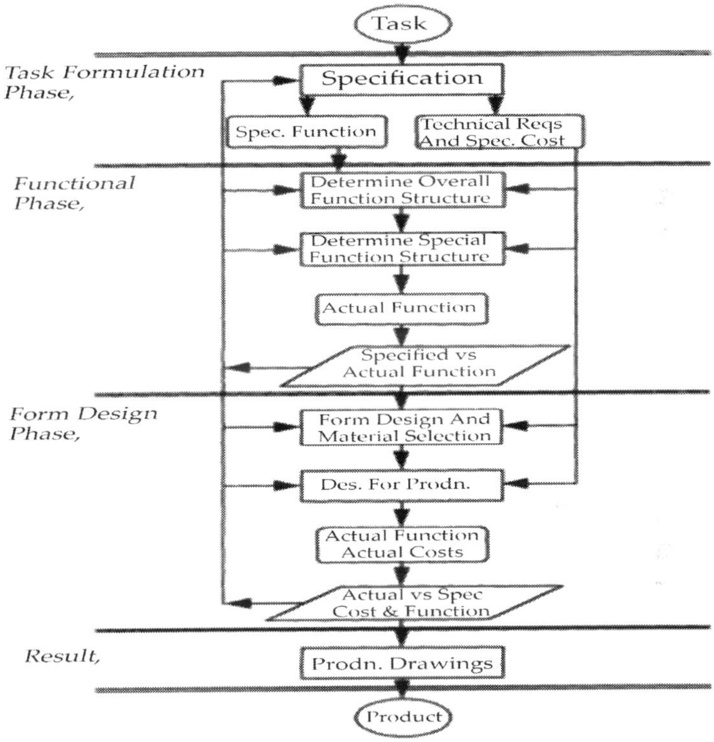

Figure 8.3: Roth's Model of the Product Life Cycle (Bessant, n.d.)

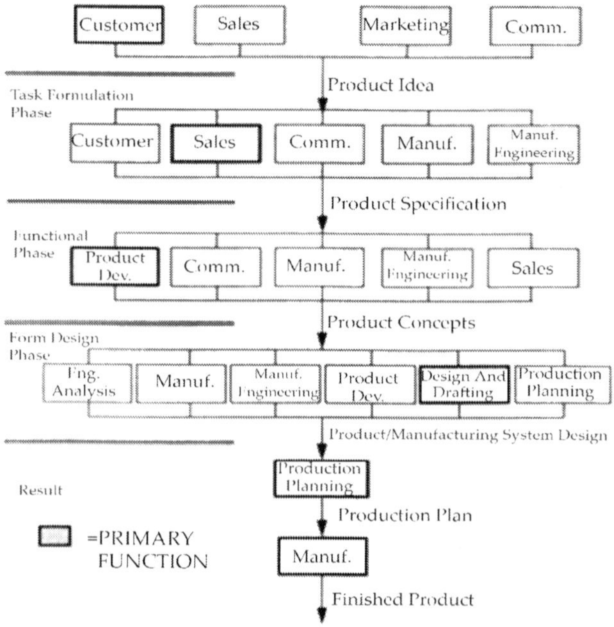

Figure 8.4: *Adaptation of Roth's Model (Bessant, n.d.)*

Figure, it can be seen that when crumbled to a level seeming useful affiliation, existing thing life-cycle models may even now be honest to goodness mechanical assemblies for the examination of C.E. We see that as the change of a thing progresses a couple of parts of the arrangement will go through the stages speedier than others, yet have kept up the key course of action of stages as the advancements which they address could be of huge worth in broaden control. A couple of limits will have an imperative part in the process at particular stages, playing out the errands for which they are chiefly careful, by then have a diminished, admonitory part at different stages.

Moreover, the model outlines the broadness of realizing which is required in such an approach, and can give an idea of the measure of the errand required in empowering the correspondence which will be required at each stage.

Benefits of Concurrent Engineering

Extending thing combination and particular disperse quality that defer the thing headway process and make it all the more difficult to foresee the impact of plan decisions on the helpfulness what's more, execution of the last thing.

- Increasing overall centered weight that results from the creating thought of reengineering.
- The necessity for snappy response to brisk changing client ask.
- The prerequisite for shorter thing life cycle.
- Large relationship with a couple of divisions wearing down working up different things in the meantime.
- New and creative advancements creating at a high rate, along these lines making the new thing be mechanical old inside a short period.

FLEXIBLE MANUFACTURING SYSTEM

A Flexible manufacturing system (FMS) is an amassing system in which there is some measure of flexibility that empowers the structure to react in case of changes, paying little respect to whether expected or unpredicted. This flexibility is overall considered to fall into two groupings, which both contain different subcategories. (Jha, n.d.)

The essential grouping, machine flexibility, covers the structure's ability to be changed to convey new thing sorts, and ability to change the demand of operations executed on an area. The worthless is called guiding versatility, which includes the ability to use distinctive machines to play out a comparative operation on an area, and also the structure's ability to hold huge scale changes, for instance, in volume, farthest point, or capacity.

Most FMS contain three central structures. The work machines which are every now and again motorized CNC machines are related by a material dealing with structure to enhance parts stream and the central control PC which controls material advancements and machine stream.

The key purposes of enthusiasm of a FMS is its high flexibility in directing manufacturing resources like time and effort with a particular true objective to make another thing. The best use of a FMS is found in the making of little game plans of things like those from a huge scale fabricating. (Jha, n.d.)

System Performance

A Flexible manufacturing system includes a PC controlled game-plans of self-ruling programmable workstations solidified with a materials dealing with system expected to make an extent of different fragment parts at low to medium volumes. The profitability of the structure is refined when the workstation set up times are low, the controlling of the parts through the system is versatile, beds are used as holding contraptions for the material dealing with system and a scope of establishments are used for part holding and region. (Jha, n.d.)

The lead components of versatile amassing systems are different assorted work pieces can be machined in the meantime, the machining game plan or, then again course through the structure can be particular for each portion part, the working game plan of the structure is not blocked by the necessity for retooling of the workstations, the materials managing and the section stream is certainly not synchronized at a relentless preset rate however is equipped to the individual machining cycles of the workstations on the course of each fragment and by and large crucially the workstations can be gotten to on an unpredictable preface. (Jha, n.d.) The era sums finished and the grouping of thing qualities make versatile bunch delivering structures appealing because:-

- Lessened set up times can be refined,
- Decreased era throughput times or lead times can be proficient,
- Lessened work ahead of time results,
- Large measures of workstation utilize are possible,
- Expanded versatility for the era of fragment parts is made available,
- Expanded response to speedy market changes is possible,
- Expanded versatility for the introduction of new things is made available,
- The structure can be made progressively.

Versatile gathering system operation is affected by the need to develop if as far as possible and organization will meet expected creation sums for a predefined thing mix while satisfying minimum target regards what's increasingly, operational levels of execution for workstation use, work ahead of time, lead times and era rates. The structure must be tweaked to satisfy a great combination of era designs in light of thing qualities moreover, the provoke advertise ask. The operational execution is affected by the fragment sums, the course to be trailed by each portion part through the system, the machining and process term at each workstation on the course, the breaking point of the vehicle units and the physical plan of the structure.

Furthermore working components fundamental to the execution of a system are the booking rules for controlling the surge of the individual parts through the structure, the nature of the collaborations between the section parts in the workstation lines and the workstation tooling limits which choose the scattering of the operations between the assorted workstations and in this manner the workstation procedure lengths.

The execution pointers for the structure can never be totally self-sufficient of each other in light of the fact that a development in the workstation utilize level is oftentimes proficient by setting up no short of what one line position on the commitment to the workstation, while the closeness of a line position on the yield side of the workstation gives additional flexibility to the vehicle unit while moving parts beginning with one workstation then onto the following. Along these lines extended workstation utilization is refined to the weakness of growing the work ahead of time, extending the get ready lead time through coating and diminishing the creation rate. (Jha, n.d.)

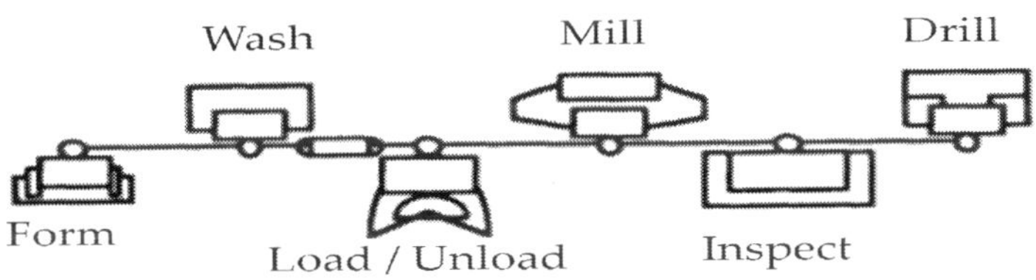

Figure.8.5: Arrangement of Flexible Manufacturing System (Bessant, n.d.)

Manufacturing Schedule

Information concerning the structure execution is required at the structure plot sort out, is required as an element of the era orchestrating limit and is fundamental while considering elective collecting frameworks. Once an era design is set up information is considered the length of the creation run, the mix of parts making up the bunch with the sums required for every special part sort, the demand of the workstations on each part course and the planning procedure spans at each workstation. A couple of systems are by then open to choose the ordinary working execution of either a current or a proposed structure.

The versatile amassing structure given in Figure 5. Is required to satisfy the era design given in Table 1. In a steady era run time of 126. 67 hours. The structure contains six workstations coordinated along a bi-directional straight track with a single transport unit. Portion spaces and completed parts are stacked and purged through a comparable workstation. The vehicle unit takes 2.5 minutes to move a segment beginning with one workstation then onto the following and one segment is carried on each bed Part stacking and exhausting at each workstation takes an advance 0.5 minutes and each workstation contains one line position at both the information and the yield ranges. The era design is portrayed by a broad assortment in the era sums, the modestly long workstation process durations required for preparing and audit appeared differently in relation to whatever remains of the procedure terms besides, the assortment in the arranged course through the system. (Jha, n.d.)

Table 8.1 Required production schedule (K Case, n.d.)

Part ID	Production Quantity	Workstation Cycle Time (minutes) Load	Mill	Form	Drill	Wash	Inspect	Unload
PI	62	15	25.0	2.0	12.0	15	18.0	1.5
P2	138	15	20.0	2.0			15.0	1.5
P3	230	15	21.0	-	5.0	15	12.5	1.5
P4	320	1.5	11.0	4.0	19.0	15	19.0	1.5

Performance Based on System Capacity

The limit investigation depends on setting up the clump fabricating time, the creation time for every part, the execution of every workstation and the use level of the vehicle unit from an assurance of the workstation process durations, the amount delivered at every workstation and the part transport time. Once the vehicle time between workstations, where each part has one less transport move then the quantity of workstations on the course and the part trade time between the workstation and the vehicle framework is known, the offices expected to deliver each part can be resolved. For each part the aggregate creation time is the whole of the individual preparing process durations and the transport time. The extent of time each part spends at a workstation and in transport is resolved as a proportion of the aggregate creation time. The working time for a workstation is figured from the aggregate of the machining process duration and the part trade times with the usage esteem communicated as the proportion of the genuine working time to the accessible booked creation time. Once the segment exchange time and the generation amounts are known the vehicle use level can be resolved from the proportion of the vehicle working time and the vehicle accessible time. In the event that the framework is intended to convey at least one sections on every bed and all parts on a bed are handled at the workstation before the bed moves to the following workstation, the normal generation amount per bed for each sort of part can be resolved alongside the quantity of beds important to meet the generation prerequisite If various diverse part sorts are in the framework at a similar time then the quantity of beds required to create each extraordinary part sort can be utilized as a measure of the normal number of parts in advance in the framework. The normal execution of the framework in endeavoring to meet the necessities of the generation plan are given in Table 2. Also, Table 3. Table 2. Demonstrates the generation timetable can undoubtedly be accomplished with the framework execution being ruled by the use levels of the processing and the investigation workstations. Noteworthy measures of sit still time are recorded for the remaining workstations. Table 3. Demonstrates the adjust of the creation time for each part is dedicated to the real assembling cycle with the vehicle action involving ten to fifteen percent of the generation time. The normal number of parts in the framework, computed from the quantity of beds required, is observed to be 4.22.

Workstation Name	Load	Mill	Form	Drill	Wash	Inspect	Unload	Trans
Processing Time (min)	1500	12075	1940	8280	9486	12516	1500	9985
Idle Time (min)	5000	925	11060	4720	3514	484	5000	3015
Utilisation (%)	23.1	92.9	14.9	63.7	73.0	96.3	23.1	76.8

Figure 8.6 Utilization levels of workstations and transport unit (K Case (ed.) n.d.)

Part ID	Total Prod Time (min)	Process Time (min)	Transport Time (min)	Process Time (%)	Transport Time (%)	Production Parts /Pallet
P1	90	75.0	15.0	83.3	16.7	144.4
P2	50	40.0	10.0	80.0	20.0	260.0
P3	69	56.5	12.5	81.9	18.1	188A
P4	83	68.0	15.0	81.9	18.1	156.6

Figure 8.7 Part production (K Case (ed.) n.d.)

Performance Based on Queuing Systems

A specific component of any adaptable assembling framework is the utilization of cradle stockpiling given by the limited line offices at the workstations or by the cushion stockpiling units in the framework. The quantity of parts in the framework and the normal time a section spends in the framework is reliant on the cradle stockpiling, the quantity of workstations, the framework design and the vehicle game plan. The framework is spoken to as a progression of free workstations where w (Beatty, n.d.)ork pieces land for preparing and if the workstation is sit without moving handling begins promptly, something else, the part joins the finish of the info line. At the point when the workstation completes the process of handling one section it either initiates preparing the initial segment in the information line or it stands inactively until the entry of the following part. The framework in this manner shapes an arrangement of couple lines in arrangement, as recommended by Hunt (1956). The principal workstation in the framework is expected to have either a steady supply of parts touching base at a settled interim of time or an arbitrary supply of parts landing about a mean entry time, Brandwajn and Jow (1988), however because of the limited line estimate at the rest of the workstations parts traveling through the framework end up plainly blocked when the information line at the following workstation is full, Berkley (1991). With limited lines between workstations the lining system turns out to be extremely troublesome and as per Papadopoulos, Heavey and Browne (1993) as the quantity of workstations in the framework is expanded it might be illogical to tackle numerically. Utilizing a solitary workstation show with a mean part entry time of 15 minutes and a vast line at the workstation the working execution of the assembling framework is given in Table 4.

Part ID	Required Quantity	Mean Process Time (min)	Mean Time in System	Mean Time in Queue	Parts in System	Parts in Queue
PI	62	75.0	1905.0	1830.0	25.0	24.0
P2	138	40.0	1752.5	1712.5	43.6	42.6
P3	230	56.5	4808.0	4751.8	84.9	83.9
P4	320	54.0	6274.5	6220.5	115.9	114.9

Figure 8.8 Performance summary (K Case (ed.) n.d.)

Performance Based on Simulation

Adaptable assembling frameworks are excessively perplexing, making it impossible to have their execution assessed without some type of PC simulation Progress in recreation innovation, the advancement of superb PC designs for liveliness and the arrangement of devoted and broadly useful reenactment programming has expanded the utilization of reproduction lately. A reproduction of the assembling framework given in Figure 1. with a consistent part landing time of 17 minutes is appeared to produces just 215 of the 750 sections required by the creation calendar of 216.67 hours. The outcomes are given in Table 5 and Table 6.

Part ID	No Requ	Parts Made	Prod Rate Pts/Hour	Total Prod Time per Part (min)	Process Time (min)	Queue Time (min)	Transport Time (min)
PI	62	55	0.25	222.2	75.0	102.2	45.0
P2	138	54	0.25	117.5	40.0	52.5	25.0
P3	230	53	0.24	155.5	56.5	69.0	30.0
P4	320	53	0.24	172.8	68.0	52.3	40.0

Figure 8.9 Production summary (K Case (ed.) n.d.)

Station Name	Proc Time (min)	Proc Time%	Queue Time%	Blocked Time%	Idle Time%	In Queue Time (min)	Out Queue Time (min)
Load/ Unload	649.5	5.0	1.7	5.1	88.2	0.50	11.85
Mill	4014.0	30.9	0.8	5.1	63.2	3.88	13.29
Form	430.0	3.3	0.6	0.0	96.1	0.25	16.07
Drill	1937.0	14.9	0.6	0.0	84.5	0.25	13.87
Wash	2430.0	183	0.6	0.0	80.7	0.25	6.40
Ins	3471.9	26.7	0.8	0.0	72.5	0.36	8.59

Figure 8.10 *workstation performance (K Case (ed.) n.d.)*

The assurance of framework execution can change extensively contingent on the strategy for assessment chose. The framework limit approach does not give adequate data about the normal framework operation and execution and the capacity to convey a predetermined item blend inside a given generation plan. The approach dependably overestimates the creation execution since it is commanded by the static limit esteems and makes no endeavor to perceive the framework progression. Lining systems and recreation endeavor to mull over the dynamic execution of the framework and perceive that powerfully the framework dependably backs off because of the distinctive parts viewing for a similar workstation assets.

Advantages

- Decreased assembling cost
- Lower cost per unit delivered,
- Greater work profitability,
- Greater machine proficiency,
- Improved quality,
- Increased framework unwavering quality,
- Reduced parts inventories,
- Adaptability to CAD/CAM operations.
- Shorter lead times
- Improved effectiveness
- Increase generation rate

Limitations

Initial set-up cost is high,

Substantial pre-arranging

Requirement of gifted work

Complicated framework

EFFICIENCY MEASUREMENT IN CUTTING EDGE MANUFACTURING SYSTEMS

Organizations occupied with make must contend in the worldwide commercial center while delivering to world-class standards. Firms utilizing Advanced Manufacturing Technologies (AMT) must react to request with reference to conveyance, value, quality and flexibility in the substance of diminishing item lifecycles and littler parcel sizes

Standard Efficiency Measurement

Measures of efficiency have been created all together that administration may practice auspicious control of generation and to improve and ease basic leadership. Generally, measures created in industry have taken cash or the most costly factor contributions as the file to be checked. Proportions of contributions to yields have been given in esteem terms with administrators consolidating various appropriate measures to accomplish ideal favorable position. The advantage of such measures lies in their utilization of accessible operational and costing information while supervisors in standard assembling operations can accomplish a level of estimation adaptability.

In innovation drove ventures a development towards higher backhanded costs, diminished direct work costs and higher capital turnover in organizations working under prohibitive market weights has decreased the criticalness of these traditional indexes. In differentiate the requirement for quantitative learning identified with assembling is a key factor. Satisfactory execution measures are expected to drive the creation frameworks and the pertinence of standard estimation frameworks which are static and frequently free of unique impact is flawed. (Alistair Barros, n.d.)

Many organizations are neglecting to distinguish factors identifying with inaccessible or insufficient cost data. Convenient administration data is expected to screen and exercise judicious business choices. On examination, existing estimation and control strategies can be identified with frameworks measuring stock, generation, quality, and costing, planning and capital use. Every one of the territories distinguished utilize either profitability or cost bookkeeping measures to demonstrate efficiency, however, most experience the ill effects of the thinness of the approach taken.

While the mix of measures may identify with control crosswise over ranges, by and by each spreads just a little region of an organization's operations, control being practiced by guaranteeing that individual segments are kept inside a foreordained arrangement. Measures are required to highlight wasteful aspects all together that healing move can be made. They should likewise be adequately conceivable to empower basic leadership and sufficiently refined to take into consideration creation and operations control. The absence of such appropriate measures in AMT has prompted discourse focused on profitability and bookkeeping strategies, each related thusly to the difficulties often experienced when justifying new advancements in money related terms.

Efficiency

Efficiency estimation for the most part depends on taking "yield" as the overseeing file. After some time, this has created as a similar measure and is regularly communicated in financial terms. The information utilized is taken from existing bookkeeping and generation frameworks, with the information lists normally picked on the grounds of corresponding impact identified with cost, traditionally, coordinate work. Nonetheless, specific elements affecting execution in high innovation make have as of now been seen to vary from those generally viewed as, most

Efficiency Measurement in Advanced Manufacture

Clearly, coordinate work. Fractional profitability measures, while remaining the prevailing way to deal with estimation, have flawed handiness. The profitability mystery as talked about by Skinner (86) has likewise turned out to be all the more broadly acknowledged and industry has encountered that endeavors to adjust contributions to standard measures, for example, machine utilization, have raised efficiency and afterward brought down it again through overproduction. (Ohno 82)

Accounting

Numerous administrators now trust that regular bookkeeping frameworks are exorbitant to regulate, neglect to control costs and typically yield mistaken information and that inside AMT, the information coming about because of standard measures can be deluding. Likewise with efficiency measures, calls for change have differed from controlling existing individual contributions to the advancement of aggregate frameworks equipped for considering factors in all cases. Brimson (88) talks about cost administration frameworks, Roth and Morse (83) suggest that quality expenses be incorporated into the hierarchical records (88), Seed (84) endeavors to supplant coordinate work and overheads utilizing a change cost while the administration of overheads, an especially risky region in AMT, is tended to utilizing action based costing. (Cooper and Kaplan 88)

In any case, all these recommended measures still identify with verifiable money related action as opposed to the regale administrators may anticipate from effectiveness measures that assistance record and control exhibit execution. Furthermore factors relating particularly to the presentation of innovation, other than the general acknowledgment that short run peripheral expenses can approach long-run minor expenses and that we should now some way or another record for the developing impact of overhead, have not been distinguished. These insufficiencies may demonstrate indispensable if the more non esteem added movement that is expected to help generation, the more noteworthy the downstream costs that are incurred. The unacceptability of customary measures lies in their being founded on the unfaltering creation of develop items with known attributes and stable innovations and endeavors to mirror the developing significance of overhead as a huge piece of aggregate cost can knot numerous critical sources of info together.

To devise measures for AMT, endeavor at outline must be made and the significance of individual factors in connection to result caught on.

Total System

The theory of Total Systems has prompted the discourse and improvement of various enhancement models that look to incorporate individual measures. Mohanty and Rastogi (86) propose the advancement of a measure to indicate Total Productivity while Son and Park (87) talk about the likelihood of quantifiable measures that may screen efficiency in three key territories, Total Quality, Total Flexibility and Total Productivity. They recommend this could be viewed as an essential assembling execution measure. Sink (85) goes ahead to recognize a

Multi-factor Approach, endeavoring to record changes after some time in critical execution regions including benefit, efficiency, cost and quality. In any case, every one of these models disregard those components that have been consider unquantifiable utilizing standard measures and recommend just that these elements must be incorporated by taking a gander at them independently and afterward permitting them impact on the basic leadership prepare. Endeavors to represent all variables utilizing complete frameworks have so far focused on the proceeded with utilization of existing efficiency and cost bookkeeping models with intermittent endeavors to incorporate components from different ranges that have been felt to now apply more impact because of the presentation of the innovation. It ought to likewise be perceived that the import of different elements will change after some time and with situation.

Capital Justification

The region of avocation was a characteristic decision for examination when identified with the high cost of AMT. The circumstance in industry has created to the point where a few supervisors have turned down recommendations creating producing innovation in light of the fact that regular strategies have demonstrated the advancement to be outlandish. Research at the University of Hertfordshire (UH) shows that somewhere else improvement has continued exclusively as a "demonstration of confidence" (Fowler et al 93). It has been said that multiple occasions a legitimate proposition has been rejected on the grounds that the readiness of the monetary examination has restricted the incorporation of essential "impalpable" variables identified with AMT. Exchange has based on endeavors to incorporate many purported indefinable advantages basic to AMT. Without a doubt, existing models neglect to assess every one of the elements that outcome from the presentation of AMT. Speculation streamlining models have been produced that endeavor to decide the impact of individual venture distribution on other money related outcomes inside the association. Be that as it may, as with other proposed proficiency measures these recommendations are either worked around existing models or endeavor to incorporate every single conceivable factor quantifiably. A need exists for thorough and compelling control over the subordinates utilized as a part of speculation examination related straightforwardly to AMT.

CHAPTER 9

Green Manufacturing

INTRODUCTION

The term "green" manufacturing can be taken in two ways: the manufacturing of "green" products, especially those utilized as a part of sustainable energy producing systems and clean innovation equipment of various sorts, and the "greening" of manufacturing — lessening contamination and waste by limiting natural assets use, recycling, reusing and what was viewed as waste, and decreasing outflows.

Green stands for environmental sustainability and comprises a wide range of concerns including, but not constrained to, air, water and land contamination, energy utilization and productivity, and waste production and reusing. Green activities plan to limit the effect of human exercises on the natural environment. (MBA, 2013)

The humanity's increasing worries for Green manufacturing can be collected into three wide groups:

Rising Emissions and Associated Climate Change

Greenhouse gas discharges have increased quickly in the current past and their production is additionally quickening. Worldwide temperatures have increased by 0.74°C throughout the last century — the fastest warming seen ever. At the present rate, emanations will twofold by 2050, contrasted with the 2000 levels. This could mean a relative temperature ascent of 4–6°C over pre–industrial levels before this current century is over. This remarkable change will gravely affect the worldwide ecosystem, hydrological system, ocean level and crop production and related exercises.

Fast Depletion of Scarce Natural Resources

With perpetually increasing populace and industrialization, the utilization of natural assets such as wood, coal, oil, water, etc. is quickly on the ascent, while their supply is contracting. This has prompted occasional confounds in demand–supply and profoundly fluctuating costs, affecting both corporate profits and customer spends. There is a swift need to (a) adequately dealing with the utilization of these assets and (b) find and create resources which are less rare such as wind and sun etc. (MBA, 2013)

Growing Waste Generation and Pollution

Increased industrialization and urbanization have resulted noteworthy increase in waste production and environmental contamination. Modern waste with chemical species can be possibly risky to life, and its dumping without treatment is accelerating land and water contamination. The release of manufacturing effluents in waterways and other water bodies is pulverizing nearby environments. As the demand and utilization of electronic products rise, e–waste is additionally turning into a noteworthy wellspring of environmental contamination.

TRANSFORMATION TO GREEN MANUFACTURING

Manufacturing firms can move on to green manufacturing by centering on three extents:

Green Energy

Green energy encompasses generation and utilization of cleaner energy. This is the first and generally self-evident step given the reliance of industry on energy. Green energy incorporates both supplying sustainable power sources like CNG, wind, sun based and biomass, and accomplishing higher energy productivity in manufacturing.

Green Products

Creating greener products is the second step in this change. 'Recycled', 'Low carbon impression', "Natural" and "organic" are getting to be noticeably prevalent business terms which are associated with Green products. Creating Green products can regularly mean higher expenses. Though, by creating Green products that are looked for by purchasers, and successfully advertising them, organizations can infer extra volumes and cost premiums, which can counterbalance their cost of green advancement.

Green Processes in Business Operations

The third challenge is installing and executing Green manufacturing processes. This involves proficient utilization of key assets, decreasing waste production through lean operations, cutting down the carbon foot–print and rationing water. Utilizing Green processes enhances operational proficiency and brings down expenses.

Forces Driving Green Manufacturing

Various organizations have begun adopting Green manufacturing as a fundamental piece of their business. These initiatives are geared by below elements:

- Rising energy and raw material costs
- Growing customer attraction for Green products
- Increasing government or administrative burdens as policy formulators present new and stricter environmental and waste administration laws
- Technological advances which open up new appealing business openings

- The need to improve competitive separation, especially for first movers or the individuals who can break the bargain between short–term higher expenses and various advantages

Green has moved from being seen as a 'vital evil' to being viewed as 'great business'. Organizations that embrace Green activities stand to be advantaged on brand improvement, political footing and administrative consistence, more noteworthy capacity to draw in and hold talent, improved client retention and potential cost investment funds. Although, these advantages require a long haul responsibility and balancing against short term goals, as the financial matters of Green manufacturing are not fully known yet

GREEN AS AN INTEGRAL PART OF BUSINESS

From the preceding decade, environmental change from Greenhouse gases has moved from being a theme of general discourse into an imperative variable adding to the financial soundness for manufacturing organizations. An imperative component of an organization's Green system is to make a Green brand that clients trust. Growing such a brand is not a convenient task and needs an incorporated advancement crusade which reliably conveys the organization's Green mission and strategies. Standard setting and product naming are capable devices utilized by organizations over various industries to fabricate validity around the organization general, as well as its products and manufacturing processes, to offer assistance to consumers in agreeing on informed choices. (MBA, 2013)

TECHNOLOGIES FOR GREEN MANUFACTURING

Today, there is a plenty of new and developing advances that guide in both, making the conventional organizations Greener, as well as making totally new ones. For instance, advancements for decreasing Greenhouse gases can be classified into five general classifications:

Carbon Sinks

This classification comprises of eminent advances in correlation with

Carbon Capture and Storage (CCS) being used in power generation plants that are ignited by petroleum derivatives such as coal. These innovations empower catching and putting away CO2 in ways such that it doesn't enter the environment. For instance, CO_2 from non-renewable energy sources is caught and put away in underground wells under heavy weight which keeps it in condensed shape.

Efficient Fuels

This classification encompasses a class of advancements that utilize cleaner fuels for power generation. such as biomass, hydal power, Integrated Gas Combined Cycle power plants etc.

Consumer Green

This includes utilizing perfect and productive fuels at the client end and arrangements encompassing demand side administration. For instance, off–grid sun oriented power applications like sunlight based water warming and building protection are incorporated into this classification.

Green Transportation

It encompasses use of green fuels in transportation. Electric vehicles, fuel cells, and bio-diesel are a few cases of this classification.

Industry Efficiency

This classification alludes to the utilization of Green production techniques and advancements in conventional ventures such as iron and steel, refining, chemicals, etc. Many of such innovations are developing in each of these businesses.

Every innovation inside these five classifications can be additionally portrayed on two measurements – development (nascent versus built up) and accessibility (nearby versus worldwide). . While a few advances such as biomass, hydro and off–grid sun oriented score higher as far as their relative technological and business development, there are others like tidal, wind and concentrated sunlight based power which are generally nascent. So also, while advances such as IGCC and CCS are universally accessible for utilization, others such as geo–thermal and waste to energy are usable just in select topographies. Grouping innovations on these two measurements gives direction in making plan of action decisions. It likewise strikes a harmony between astutely bridling the income pools of today and proactively situating for catching future income pools. Contingent upon their craving for asset duty, organizations can settle on a very basic level diverse decisions. Players resolved to long haul reception of Green can choose an arrangement of advancements to actualize – going from develop and promptly accessible ones to making specific wagers in new ones. (MBA, 2013)

Correspondingly, many develop as well as rising Green advancements are accessible in the areas of water and waste treatment.

GREEN ENGINEERING

Green engineering deals with the design and plan of products and processes by applying monetarily and technically viable processes and products in a way that at the same time decreases the measure of contamination that is created by a source, limits confrontation to potential dangers as well as securing human wellbeing without giving up the financial effectiveness and viability. As such, green engineering is not really a engineering paradigm in itself, but rather an all-encompassing designing system for all design genres of engineering.

Principles of Green Engineering

Green engineering sticks to nine controlling standards. A product designer must endeavor to:

- Design and engineer processes and products comprehensively, utilize systems examination, and incorporate environmental effect assessment instruments.
- Conserve and enhance regular ecosystems while ensuring human wellbeing and prosperity.
- Employ life-cycle approach in all design and engineering exercises.
- Ensure that all material and energy data sources and yields are as characteristically safe and considerate as possible.
- Minimize exhaustion of natural assets.
- Try best to counteract waste.

- Develop and apply engineering systems, while being conscious of nearby environment, aspirations, and societies.
- Create engineering solutions past present or predominant technologies; enhance, advance, and develop to accomplish supportability.
- Actively draw in stake holders for development of engineering works.

SYSTEMS APPROACH TO GREEN MANUFACTURING

To differing degrees, all engineering disciplines take part in green engineering. This incorporates economical design, life cycle investigation, contamination avoidance, plan for the environment, design for dismantling, and design for reuse. As such, green engineering is a sub-decipline of sustainable engineering. Green engineering encompasses four basic ways to deal with challenges of enhancing processes and products to make them more proficient from an environmental standpoint.

Waste decrease

Materials administration and management

Pollution avoidance

Product improvement.

Green engineering approaches product design from a systematic point of view, which implies that various expert orders must be intermingled. Along with all engineering disciplines, green engineering incorporates land utilization management, design, landscape engineering, and other architecture fields, as well as the social sciences(to determine how different individuals utilize products and services. Engineers and designers have dependably been worried about space. Engineers consider the availability of space. Designers see the site map as an arrangement of fluxes over the limit. Planners consider the blends of these systems over bigger areas. The life cycle examination is a vital green engineering device, which gives an all encompassing perspective of the totality of a product, process or movement, encompassing crude materials, manufacturing, transportation, dispersion, utilization, upkeep, reusing, and final disposal. As such, assessing its life cycle should yield a whole picture of the product. The initial phase in a life cycle assessment is to accumulate information on the stream of a material through the society. Once the amounts of different segments of such a stream are known, the vital capacities and effects of each progression in the production, make, utilize, and recuperation/transfer are evaluated. Subsequently, in practical plan, engineers must streamline for factors that give the best execution. (MBA, 2013)

The systems approach utilized in green engineering is like value engineering. Daniel A. Vallero views green engineering as a type of value engineering on the grounds that both systems require that all components and linkages inside the project be considered to upgrade the value of the product. Each segment and step of the system must be tested. Ascertaining general overall value is not only be a project's cost-adequacy, as well as different other esteems, including environmental and general wellbeing elements. Therefore, on the wider scale value engineering is like and can be indistinguishable to green engineering, since value engineering is aimed for viability, not simply productivity, i.e. a project is intended to accomplish numerous destinations, without giving up any vital esteems.

Efficiency is an engineering term for the proportion of a contribution to a yield of energy and mass inside a system. As the proportion approaches 100%, the system turns out to be more productive. Adequacy requires that efficiencies be met for every segment, additionally that the mix of parts prompt a successful, multi value based design. Green engineering is likewise a kind of concurrent engineering, since tasks must be parallelized to accomplish various plan targets.

FOUR PILLARS OF GREEN TECHNOLOGY POLICY

- Energy – Strive to achieve energy independence and encourage effective and efficient use of energy.

  An imperative mainstay of green technology is energy production, which looks for new inventive approaches to produce helpful energy from futile waste, as well as to develop new technologies to augment the harvesting of energy. Energy productivity has turned out to be a financially savvy plan for energy economies by limiting the amount of energy utilized.

- Environment - Safeguard and limit the effects on the environment.
- All human exercises have an effect on the environment. That's why, the relative soundness of the environment will decide and will add to the nature and size of exercises in the other pillars of this green technology policy. We have noticed the major worldwide environmental difficulties: environmental change depleting nonrenewable assets, contracting regular natural surroundings, decreasing biodiversity, sea degradation; and increasing human populace weights. Obviously, all of these issues have monetary, social and social component (MBA, 2013)
- Economy - Enhance the national financial advancement using innovative technologies. It is centered around the fascination of new organizations and individuals to green technology. This is basic to the city's development and manageability and assists us in building a solid and dynamic community. The hiring of capable individuals in public, private, and non□ profit divisions is essential requirement to accomplishing this. Existing organizations and the occupations they make are fundamental parts of a solid, supportable economy. Together we can accomplish and be the distinction that has the effect.
- Social - Enhance the personal satisfaction for all.
- The Social Equity Pillar will help social organizations and occupants to bring issues to light about social needs and to draw in both residents and social partners to plan and act in light of these necessities. The final goal will be to enhance the prosperity of the entire society. Together, we will manufacture social capital in the group amongst people and gatherings with a specific end goal to empower collective activity on tasks of normal intrigue. With this, we push the significance of individual and amass prosperity and security, including full access to to social health insurance, food, residence, and education these being the basic parts for full involvement in social, environmental, and financial exercises.

SECONDARY RECYCLING UTILIZATION AND CORE TECHNOLOGY

The objective of secondary assets use is to utilize assets from diverse sources and in different structures, make new products with low energy utilization, spare assets, avoid contamination, make practical commitment to the central economy, and approach and understand the objective of setting up asset sparing society. The prompt objectives are: accomplishing essential far reaching use of secondary assets; increasing the reuse proportion of metal to more than 80%; setting up a total technology system for the use of fumes, natural/inorganic mixes, and organic scraps; and forming mechanical chain for secondary assets reusing.

In the following 20~50 years, there will be three tasks for clean and productive usage of secondary assets. First, to create effective and clean technological advances for the usage of slag and tailings produced from mining and metal dressing, as well as metallurgy slag and efluent produced from metallurgical processes, accomplishing abbreviated green division of different metals and getting high-esteem products. Secondly, to create reusing technologies, for non-ferrous metals, waste mechanical and electrical products, and E-wastes, accomplishing effective seperation and reusing of multi-parts materials. Containing metal, natural, and inorganic substances. Thirdly, to create advances for the propelled use of the multi part assets and for the financially savvy improvement of high-esteem parts with low focus, accomplishing high-esteem use of organic deposits and waste gasses. It is normal that in the following 20~50 years leaps forward will be made in key technologies for perfect and effective use and reusing of secondary assets. (Tianran Wang, 2012)

Key Technologies

The reusing and recycling technology of secondary assets is a far reaching innovation incorporating seperation science and technologies, catalytic technologies, natural technologies and innovations and process addition/ integration technologies. The investigate targets are factory and family unit wastes as gas, liquid, and solid. The research objective is to isolate the valuable parts of the materials to make them into high-value assets or change them into new products under specific conditions. The environmental technological innovations emphasize avoiding from poisonous products or decreasing the aggregate sum of contamination production and their environmental effects by creating decomposable contaminations with the use of environmental self-purifcation capacity in manufacturing process and secondary assets recycling process. Because of the complications of secondary assets and environment issues, the broad recycling and using of the considerable number of wastes such as the separation and investigation of the entire components, the reusing of trace elements, the minimal effort detoxification, and etc., still have many difficulties; in any case, leaps forward are expected to be made in the key advancements as stated below. (Tianran Wang, 2012)

Multi-Metal Shortened Green Separation and Recovery Technology

Scrap metal, metal alloys, tailings, metallurgical process residues, muck, dust and other secondary assets by and large contain different metals. The savvy programmed sorting, productive metal refining, new alloy leaching, composite particular filtering, effective new extractants, and microbial particular obsession technologies are making advancements. For example, vanadium and chromium clean separation technology of factory carcinogenic metallic slag; innovative separation technology of vanadium, titanium, and iron in vanadium steel slag and high titanium-bearing slag; and very proficient partition and waste recuperation technology and equipment for lethal heavy metals, nonreactive or noble metals, and rare earth metals in multi-component factory wastes are accomplishing specialized achievements and modern applications. The entire flow mixing and optimization, the zero waste release of the tailing, and waste recuperation process are being accomplished. (Eric M. Taleff. St, 2004)

Sulfur Recovery Technology in Metallurgical Industry

Sulfur containing gas and fluid wastes are fundamental toxins in manufacturing and metallurgical processes, and the essential way to deal with dispose of their environmental effects is sulfur reusing. The present day products of sulfur on the planet predominantly originate from the natural sulfur, corrosive acid gas, acrid raw petroleum, pyrite, and non-ferrous metal sulfides; while the measure of the recuperation sulfur will turn out to be increasingly bigger part of sulfur assets later on. In the following 20~30 years, efficient desulfurization processes and new bio-desulfurization advancements will assume vital parts, and in the following 30~50 years over 85% of the sulfur supply on the planet will originate from the recuperation sulfur. (Tianran Wang, 2012)

The Recycling Technology of Waste Gas

In the following 20~50 years, noteworthy achievements will be made in the field of green separation technologies and chemical cleaning technology for use of factory waste gasses including SOx, COx, or CxHy. Due to the fruitful utilization of proficient separation media, exceedingly efficient chemical treatment, and natural catalysts, utilizing waste gas to produce sulfur and sulfuric acid straightforwardly is going to emerge as another industry, acknowledging zero waste gas release. (Tianran Wang, 2012)

The Recycling Technology of Waste Water

In the following years, the advances of utilizing low concentration metallic ions in waste water to get high-esteem products, new materials for membrane refining, setup for membrane refining, and exceptionally specific novel ion particle exchange resin and green products are accomplishing leaps forward. By reusing water asset completely and using solute to acquire high esteem products, waste water zero release can be figured out. (Tianran Wang, 2012)

The Technology of High Value Utilization of Biomass Residue

Mostly, other than sugar and trace measures of inorganic segments, biomass deposit basically comprises of lignin, fibers and protein. Green technology advances such as steam blast, high catalyst cracking processes, protein hydrolysis, and novel ionic fluid media for using natural deposit to get biodiesel, methanol, ethanol, glucose, protein, and cellulose subsidiaries are acquiring leaps forward and getting wide modern applications in the following years. (Tianran Wang, 2012)

Waste Recovery of Waste Organic -Inorganic Composite

Organic- inorganic composite waste is a new type of waste originated from modern industries, existing widely in the pCBs, medicines, fertilizers, catalysts, membrane materials, and other new materials. In the next 20 years, the low cost green separation technology for the separation of organic and inorganic components from composites is expected to achieve a breakthrough, and in the next 30 years, high-value and the comprehensive utilization of waste organic inorganic composites, PCBs as representing, technology is expected to realize industrialization. (Tianran Wang, 2012)

Optimized Integration of Whole Tailing and Slag Treatment Process

Process enhancement is equipped for adjusting materials, water, and energy exactly, and energy preservation and utilization decrease can be seen by improving and incorporating the entire tailings and slag processing unit. The expansion and advancement of the whole process technologies covering tailings slime gathering, pre-treatment, partition, cleaning / purification, and reusing technology will be completely finished in coming years on the basis of improvement in every single operation. (Tianran Wang, 2012)

Industry Chain of Secondary Resources Recovering

Use of only secondary assets is frequently costly and may create new waste products, producing more contamination in the environment. Through exhaustive format and natural resource planning, the business chain for secondary assets recouping and accomplish flexible joining of upstream and downstream is setting up. Business chain of secondary assets recouping and the biological modern industrial park for using of all segments is being built up. (Tianran Wang, 2012)

Green Process System Integration Technology

Begun in about 1970s, process system integration technology has been created to limit energy utilization by upgrading and coordinating specialized processes all in all. Green process system reconciliation / integration technology is the incorporation of information and computer technology with process industry, and its emerging direction is multi-scale green process system integration. Taking materials-information-energy flow of the system as research goals, green process system combination technology contemplates how this sort of system takes care of the demands for ideal economy improvement and environment assurance all the while by assets designation, scheduling, administration, design, and operation control. The goal is to accomplish the streamlining of practical production on technology and structure. The scope of research has secured many levels such as chemical molecule designing, optimizing the process, and biological industry park system integration. The key emerging advancements include molecule design and product engineering; smaller scale advancement and integration; process optimization and intelligent equipment;process reconciliation for huge complex systems. (Tianran Wang, 2012)

High-Performance & Low Energy Consumption Pretreatment Technology

Pretreatment is the beginning step of biomass handling, and it is a consistent work to flourish superior and low energy utilizing pretreatment methods for various biomass types and conversion technologies. The pretreatment of straw and other lignocellulose has been contemplated widely, and acid pretreatment and steam blast pretreatment have become commercial. In light of the serious necessities for the machinery and substantial contamination, acid pretreatment process is slowly turned down. Steam blast pretreatment process is a perfect technology because of less contamination and low energy utilization. Though it should be enhanced in areas such as; making it a continuous steam blast technology, decreasing energy utilization, and moving forward the efficiency of the system. Microwave technology and radiation advances are likewise step by step used for pretreatment of straw. It is being predicted that energy utilization in straw pretreatment can be decreased by more than a half in coming years. (Tianran Wang, 2012)

High-Performance & Low-Cost Cellulase Production Technology

The increased price of cellulose is the significant snag for the organic transformation of lignocellulose. Let's consider cellulosic ethanol production, it is accounted that the base level costs of cellulose hydrolysis can be set at 10-20 pennies/gallon of the ethyl alcohol, on the other hand, the starch-based ethanol costs just 1-2 pennies/gallon. It is thus imperative to decrease the cost of enzymes for cellulase that transform lignocellulose materials into microbial goods. It is predicted that the price of enzymes for cellulase will be decreased significantly through screening and remaking of strains and production technology developments in the coming years. (Tianran Wang, 2012)

New Media of Ionic Liquids Applied in Cellulose for High-Value Product Manufacturing Technology

There are much energy intensive, complex processes and harsh environmental contamination issues in conventional cellulose handling technology. The fundamental issues for cellulose substance conversion are growing new non contaminating and green solvents. It will be an extraordinary breakthrough however to conventional method for cellulose-base production by using cellulose diffluent ionic fluid and non- volatile fluid as the media for delivering functionally recovered cellulose and useful cellulose base derivatives. It is predicted that the ionic fluid cleaner technology arranged for making cellulose materials is going to be industrialized. (Tianran Wang, 2012)

Modern Biotechnology

Separating the high quality strains is the fundamental to organic-conversion. Through the separation and preparation of strains, joining present day hereditary / genetic engineering, metabolic engineering, and genomics advances, solid resistibility strains which are adjusted to deliver target product from the lignocellulose will be developed and separated. A yeast is going to be developed that will utilize sugar of five and six carbon atoms for creating ethanol. (Tianran Wang, 2012)

Lignin Modifying and Processing Technology

Lignin belongs to a sort of cellulose fragrant polymers, and the amount of which is simply lower than the natural cellulose. Just like the naturally found random copolymer phenols, it is built by benzene propane by ether bond and C–C bond associated together with a 3-D body structure. Consistently lignin quantities of around 60 billion tons can be created all inclusive in an year. With numerous different sorts of dynamic functional groups, lignin is accepted to be a incredible green substance material because of its prevalent properties, such as inexhaustible, biodegradable, nontoxic, and minima! effort. As of now lignin handling technology did not accomplish considerable advance, most lignin was utilized as fuel or disposed of. In the following couple of decades lignin handling technology is going to be essentially enhanced, and will leap forward in mixing lignin with phenolic base resins, polyurethane, polyester, polyether, soy protein, and other natural substances by means of compound response and physical mixing to create exceptional useful polymer materials. (Tianran Wang, 2012)

Bio-refnery Technologies for Mass Production of Chemical through Non-Petroleum Route

The manufacture of substances like ethylene, propylene, and other basic crude products are intensely subject to oil assets. With the lack of oil assets, supplanting of oil refinery with bio-refinery is an unavoidable demand for social advancement. By a progression of oil based innovations such as polymerization, cracking, hydrogenation, and aromatization chemicals will be created. It is predicted that by 2050, 45% of the fossil chemicals will be supplanted by bio products. (Tianran Wang, 2012)

Technology Integration of Biomass Comprehensive Utilization and the Establishment of Ecological Industry Chain

Contrasts of biomass materials, as well as assorted qualities of technical advances and different products make the bio transformation process a convoluted engineering system. It needs to grow new innovations in different fields and incorporate different innovations effectively to shape biological industry chains; to deliver financial, social, and natural advantages. Truly, the improvement and use of biomass assets includes many controls and numerous ventures, and their advancement and use does not mean a putting together the basic parts following a portion of the above mentioned technologies and processes, but it means setting up a multi-disciplinary system of sciences, advances, and productions. (Tianran Wang, 2012)The procedure of biomass transformation reconciliation technology takes after these standards

- Using and depleting of every segment of thc biomass materials on the basis of multi-layer grouping strategy. As indicated by the attributes of parts of plant, outline ideal production course, permitting every segment to have the capacity to change into the fit and final product.
- The most reduced energy rule. Based on the plan, design or forming of machinery, mix of heat and energy usage and different strategies to amplify energy use effciency.
- Developing unique and creative technology by coordinating different old and innovative technologies. Shape the environmental chain, the final design for biomass assets advancement, and it is predicted that, for starters environmental chain of biomass assets will be built up. The improvement of these key innovations is the target, the fundamental target products are bioethanol, other bio chemicals and bio-degradable plastics of many types.

Cascade Conversion Technologies for High-Value Integrated Utilization of Coal

At current, burning is the principle approach to expend coal, representing around 85%. Coal is the one of a kind fossil assets containing a huge amount of hydrocarbon structures. Present day burning and other industrial gasification advancements would totally wreck coal structure and subsequently can't accomplish full usage of the substance materials including basic aromatic hydrocarbon molecules which are till date produced by oil refining processes. However, the lack of oil assets asks the improvement of other production courses for aromatic hydrocarbon material. The cascade coal transformation and high-esteem coal use advancements highlighting the usage of aromatic hydrocarbon molecules and vivacious attributes of coal to accomplish multi-production of chemicals and power, as well as the usage of the natural structure and the C/H components of coal to accomplish multi-production of fragrant / aromatic hydrocarbon chemicals and amalgamation gas, reflect the key need in the region of energy assets. In this field, world is settled a strong establishment and with advance endeavors in the following years, innovations and methods for the sweet-smelling aromatic hydrocarbon molecules and warm/electric poly generation as well as the sweet-smelling hydrocarbon and synthesis gas poly generation will be shaped and the industrialized. (Tianran Wang, 2012)

GREEN AWARENESS

Green manufacturing is essential, not because of tight rules and controls or money saving advantages, however additionally on the grounds that clients are demanding it. In a latest survey of clients in many countries of different economy levels, around two–thirds of the members communicated the conviction that the environment is in a poor shape and that environmental issues are an essential danger to the general public.

Not exclusively are clients becoming increasingly mindful and familiar with Green, they are moreover receiving Green propensities and purchasing Green products. The proceeding with development of Green cognizance around the globe introduces an enormous open door for brilliant organizations. As indicated by the survey discoveries, while buyers trust that as people, they can and ought to add to supportability by receiving Green products, they additionally hold organizations to a higher standard with regards to being Green.

The term Green is perceived the world over as shorthand for environmental cognizance. However, when asked to characterize Green and their desire from Green products, clients had a scope of reactions relying upon where they lived and the kind of products they purchased. It is, consequently, basic for organizations to find how their objective buyers feel about Green, what they anticipate from Green products and what costs they will pay for them. Another finding of the survey was that around half of these clients purchased Green products. The survey likewise demonstrated that purchasers significantly esteem the immediate advantages that Green products offer, such as – unrivaled freshness and taste, the guarantee of security and wellbeing, and saving funds on energy costs. They will pay higher costs for Green products that have better quality discernment. While looking for Green is getting to be plainly regular in numerous nations, shopping propensities differ significantly by product classification. Certain Green product classifications like paper, sustenance products, expendable home products, customer durables and magnificence products are more mainstream than others and are purchased all the more frequently.

BARRIERS TO HIGHER GREEN CONSUMPTION

As said before, numerous clients, especially in developed and progressed nations, will pay a higher value for Green. Their eagerness to pay more relies upon a product's classification and seen benefits, and is most astounding for sustenance and client durables. The discoveries of the survey show obviously that cost is not a critical impediment for some purchasers. Truely, value positions much lower as a constraint to Green purchasing than absence of consciousness of Green choices or a seen absence of decision. (MBA, 2013)Clearly, mindfulness is a basic lever for increasing offers of Green products. It is assessed that organizations lose, on a normal, almost 20 percent of potential purchasers when buyers are not satisfactorily educated about the manageability aspect of their offerings. Organizations need to deliberately design and put resources into their client mindfulness projects and work with their retailers to give satisfactory rack space and perceivability to guarantee their Green endeavors are completely utilized.

FRAMEWORK FOR ADOPTING GREEN MANUFACTURING

Challenges in Adopting Green

Indeed, in intense economic situations, the process for adopting Green stays convincing. There is more noteworthy acknowledgment of the goals of getting to be Green and understanding that Green has to address every one of the three areas –

- Green energy
- Green products and
- Green processes

Organizations confront challenges on different fronts, most fundamentally in giving administration to such an exertion. Organizations need to move from

- Approaching Green as restricted, frequently confined activities with limited focus to a more comprehensive approach,
- Meeting administrative consistence for creating eco–advantage, and
- Viewing activities as cost focused for assessing them as business open channels.

This requires a noteworthy change which to succeed, requires a systematic approach and a structure tending to the three chief obstacles to unequivocal activity:

- Companies don't completely understand drivers and issues important to them and their businesses, and what manageability intends to them.
- Companies confront challenges in demonstrating the business case – or notwithstanding finding a convincing case – for maintainability. The activities are not a need for most, and frequently the financial aspects are not surely known as innovations and expenses are as yet advancing.

Even the organizations that embrace Green activities play out these exercises as fringe to their center business and not incorporate into their corporate system. Thus the execution is defective and they have failure in understanding the full advantages

Financial Assessment and Making Strategic Choices

Like any major transformational work, achievement in receiving Green expects organizations to understand the full arrangement of actualities on expenses and benefits, and the whole scope of Green measures accessible to them. When this reality base is produced, organizations need to choose their Green activities based on both, financial and vital assessments of the decisions they recognize. A financial assessment requires evaluating the "esteem" created over the long haul through these activities. It should cover all drivers of significant worth creation – from quantitative measurements like evaluating force and cost reserve funds, to subjective ones like representative enrollment and engagement – something else, a portion of the long haul advantages of Green won't get caught as a feature of the business case. (MBA, 2013)

Performing a monetary assessment is just a single piece of the story. Having put forth a feasible defense, organizations need to settle on a key decision on how Green they need to be, and why. The decision of activities could fluctuate depending not just on the hidden financial matters of the choices, additionally available setting and open doors for vital separation. Possibly, organizations can be:

- Planet aloof: where the measures embraced are negligible,
- Good residents: where they complete a couple of disconnected activities which are the absolute minimum that clients demand or controllers stipulate, or
- Green trailblazers: where they attempt to remain on the ball on manageability and change the issue from one of hazard administration to that of top line development and a key business opportunity

While the initial two decisions don't enable the organization to completely use the capability of Green and are just applicable with a transient focal point, the third confers the organization to a thorough Green methodology and to taking full advantage of the activities.

EXECUTION FRAMEWORK

Getting to be noticeably Green is a long excursion of change. To succeed, satisfactory consideration is required on arranging and execution of the activities. Early wins and victories are critical to build force. It requires a completely dedicated top administration, tight occasional audits and consistent inside and outer correspondence.

A basic three–step usage structure can be taken after covering each of the three areas of activity – Green energy, Green products and Green processes.

Plan

Green activities must be calculated into the business technique, future asset scheduling and planning works. For instance, organizations need to design thoroughly to increase the utilization of Green energy, move the product portfolio to Green products and update business operations towards Green processes. A manageability sanction, based on here and now and long term objectives, must be laid out with Green targets and measurements. Organizations ought to create

Green records or scorecards measuring the effect of the Green activities they have attempted, set particular focuses on those files and track advance against those objectives.

Execute

With a vigorous arrangement set up and targets obviously characterized and observed, Green should be coordinated over the esteem chain and made a piece of the center business.

- Green energy: Manufacturing organizations with high energy utilization need to move towards utilizing cleaner energy and plan for increasing the productivity of its utilization. Setting up hostage wind or sun based power generating units and utilizing energy productive practices, such as introducing LED lighting or better utilization of sunlight in building configuration, can go far towards lessening the energy power of operations.
- Green products: To move towards a Green product portfolio, organizations should direct an assessment of their products based on

  (a) how Green are the assets and energy being utilized.

  (b) how Green is the product amid the lifecycle of its utilization.

  (c) how Green is the manufacturing procedure.

By evaluating these parameters, organizations can display the Green estimation of their product in advertising. In the arranging stage itself, organizations should set out focuses for this metric, and then occasionally assess advance against these objectives.

- Green processes in business operations: Companies need to bit by bit upgrade business processes utilized as a part of various parts of the esteem chain. This could incorporate moving to more supportable manufacturing alternatives, rolling out improvements towards decreasing waste, increasing reusing,

reusing assets and boosting all providers, channels, clients and workers to receive comparative measures.

Communicate

Alongside well thought usage, a very much defined advancement battle for Green activities is similarly essential to completely use their potential advantages. Client training efforts about Green product offerings and the Green introduction of the firm in terms of energy and processes, can convert into increasing incomes.

ADVANTAGES AND DISADVANTAGES OF GREEN MANUFACTURING

At the point when a business settles on the choice to wind up noticeably a green, or environmentally well disposed, producer, it intentionally chooses to advance certain esteems. These qualities fixate on the objective of securing the environment yet can likewise concentrate on things such as industrial advancement and advantage.

Profiting the Environment

Green manufacturing can specifically profit the environment. For instance, green manufacturing can diminish waste and hurtful emanations and work toward protecting assets that are limited and nonrenewable. Numerous clients need to help organizations that execute green manufacturing, so by actualizing this procedure, a business can likewise increase new clients.

Profiting the Business

In light of general society's agreement about the need to protect the environment, an organization can specifically enhance its advertising by executing green manufacturing. Furthermore, this procedure can bring down expenses for the business over the long haul through the usage of more proficient systems and cultivating an organization culture dedicated to development in processes. Further, these more proficient processes can bring down the measure of waste a business produces.

Surviving the Transition

One of the burdens of green manufacturing a business encounter is the move to green manufacturing. Organizations should find subsidizing sources to back the change to green manufacturing. In spite of the fact that the procedure ought to at last spare the business cash, in the here and now executing green manufacturing can cost a business noteworthy measures of cash. Further, the move by and large requires the execution of new manufacturing processes, as well as the capacity to plan and assemble the important technology and machinery to do green manufacturing. Also, a business will regularly need to discover new ability to come in and instruct current representatives on the most proficient method to work in the new green manufacturing environment.

Worldwide Trade Disadvantages

Another potential detriment to making strides toward environmental friendliness identifies with universal trade. Numerous organizations waver to actualize a green manufacturing system since they expect that this procedure will turn into an obstacle to the free stream of merchandise and gainful exchange bargains. This delay can apply to an extensive variety of made products and enterprises. For instance, a business may falter to actualize green manufacturing since it fears losing a global provider that does not take an interest in green manufacturing.

References

1. Alistair Barros, A.G.E.K., ed., n.d. Business Process Management: 10th International Conference. Springer Shop.
2. Anon., n.d. [Online].
3. Anon., n.d. americanhydroformers. [Online] Available at: http://www.americanhydroformers.com/category/what-is-sheet-hydroforming/ [Accessed 6 July 2017].
4. Anon., n.d. http://mechanicaluniversity.blogspot.com. [Online] Available at: http://mechanicaluniversity.blogspot.com/2014/07/what-is-super-plastic-forming.html [Accessed 10 July 2017].
5. Anon., n.d. https://www.researchgate.net/. [Online] Available at: https://www.researchgate.net/figure/270396870_fig1_Figure-1-Schematic-of-forward-and-backward-tube-spinning-show-the-relative-flow-of [Accessed 7 July 2017].
6. Anon., n.d. hyperphysics. [Online] Available at: http://hyperphysics.phy-astr.gsu.edu/hbase/thermo/electrol.html [Accessed 5 July 2017].

7. Anon., n.d. M&DC Purchasing & Supply Chain: Material Management. [Online] Available at: http://mdcegypt.com/pages/purchasing/material%20management/Production%20Activity%20Control.asp [Accessed 10 July 20117].
8. Anon., n.d. mechanicalengineeringblog. [Online] Available at: http://www.mechanicalengineeringblog.com/4772-electro-chemical-grinding-non-traditional-machining-process/ [Accessed 6 July 2017].
9. Anon., n.d. novelmech.com. [Online] Available at: http://novelmech.com/2016/05/electrochemical-grinding-process.html.
10. Anon., n.d. nptel. [Online] Available at: http://nptel.ac.in/courses/112107144/Metal%20Forming%20&%20Powder%20metallurgy/lecture9/lecture9.htm [Accessed 10 July 2017].
11. Anon., n.d. princeton. [Online] Available at: https://www.princeton.edu/~maelabs/mae324/glos324/superplasticforming.htm [Accessed 10 July 2017].
12. Anon., n.d. quia. [Online] Available at: https://www.quia.com/jg/1961543list.html [Accessed 7 October 2017].
13. Anon., n.d. researchgate. [Online] Available at: https://www.researchgate.net/publication/229510367_Production_Processes_and_Equipment [Accessed 4 July 2017].
14. Anon., n.d. Scientific & Academic Publishing. [Online] Available at: http://article.sapub.org/10.5923.j.ijmee.20120104.03.html [Accessed 5 July 2017].
15. Anon., n.d. substech. [Online] Available at: http://www.substech.com/dokuwiki/doku.php?id=squeeze_casting [Accessed 5 July 2017].
16. Anon., n.d. thelibraryofmanufacturing. [Online] Available at: http://thelibraryofmanufacturing.com/high_energy_rate_forming.html [Accessed 7 July 2017].
17. Anon., n.d. www.inoxveneta.it. [Online] Available at: http://www.inoxveneta.it/en/tube-hydroforming/ [Accessed 3 July 2017].
18. Beatty, C.A., n.d. Tactics for the implementation of advanced manufacturing technologies. AbeBooks.
19. Benedict, G.F., 1987. Nontraditional Manufacturing Processes.

CRC Press.

20. Bessant, J.R., n.d. Managing advanced manufacturing technology: the challenge of the fifth wave. Blackwell Publishing Limited.
21. Black, B.J., 2015. Workshop Processes, Practices and Materials. Routledge.
22. Bralla, J.G., 2007. Handbook of Manufacturing Processes: How Products, Components and Materials are Made. Industrial Press.
23. Chee Kai Chua, K.F.L., 2014. 3D Printing and Additive Manufacturing: Principles and Applications. World Scientific.
24. Company, D.P., 2004. 21st Centuty Manufacturing. DIANE Publishing Company.
25. Daniela Lohwasser, Z.C., ed., 2009. Friction Stir Welding: From Basics to Applications. Elsevier.
26. David Ian Wimpenny, P.M.P.L.J.K., ed., 2016. Advances in 3D Printing & Additive Manufacturing Technologies. Springer.
27. Davis, J.R., ed., 1995. ASM Specialty Handbook: Tool Materials. ASM International.
28. Altintas, Y., Munasinghe, W. K. "A Hierarchical Open-Architecture CNC System for Machine Tools", Annals of CIRP, Vol. 43/1, pp. 349–354, 1994.
29. Altintas, Y., Erol, N. A. "Open Architecture Modular Tool Kit for Motion and Machining
30. Process Control", Annals of CIRP, Vol. 47/1, pp. 295–300, 1998.
31. Aronsoon, R. B. "Machine tool 101: part 1, trends", Manufacturing engineering, pp. 31–36,
32. January 1994.
33. Aronsoon, R. B. "Machine tool 101: part 5, controls", Manufacturing engineering, pp. 53–57,May 1994.
34. Astr¨ ° om, K. J., H¨ agglund, T. "Automatic Tuning of Simple Regulators with Specifications on
35. Phase and Amplitude Margins", Automatica, Vol. 20, No. 5, pp. 646–651, 1984.
36. 6. Auslander, D. M., Tham, C. H. "Real-Time Software for Control", Prentice Hall, Upper Saddle River, NJ, 1990.

37. Babb, M. "IEC 1131-3: A Standard Programming Resource for PLCs", Control Engineering,pp. 67–76, February 1996.
38. Bollinger, J. G., Duffie, N. A. "Computer Control of Machines and Processes", Addison Wesley, Boston, MA., 1989.
39. Brecher, C., Voss, M., Almeida, C. "Component-based Open Control Systems using Open
40. Source Software", Proceedings Third CIRP International Conference on Reconfigurable Manufacturing Systems, May 2005.
41. Brouer, N., Weck, M. "Feature-Oriented Programming Interface of an Autonomous Production Cell", Fourth IFAC Workshop, pp. 223–228, 1997.
42. Chang, C. H., Melkanoff, M. A. "NC Machine Programming and Software Design", Prentice Hall, Upper Saddle River, NJ., 1989.
43. Chang, H., Lee, K. "Design and implementation of real time multitasking kernel under DOS
44. environment", Journal of Korean Automatic Control and System Engineering, Vol. 3, No. 4, pp. 373–380, 1997.
45. Choi, I., Suh, S., Kim, K., Song, M., Jang, M., Lee, B. "Development process and data management of TurnSTEP: a STEP-compliant CNC system for turning", International Journal of Computer Integrated Manufacturing, Vol. 19, No. 6, pp. 546–558, 2006.
46. Chon, D., Kim, H. "Generating velocity profile for NURBS curve by look-ahead method", Proceedings Fall Conference of KSPE, pp. 977–980, 1997.
47. Chung, C., Lee, K., Lee, B. "Software design and implementation for robot controller based on real time operating system", Proceedings Conference of KACC, pp. 1246–1251, 1994.
48. Chu, J., Lee, H., Lee, Y., Jun, K. "A new contour error modeling method for cross coupling control", Journal of Control, Automation and System Engineering, Vol. 3, No. 4, pp. 389– 397, 1997.
49. Comer, D., Fossum, T. V. "Operating System Design, Volume I The XINU Approach", Prentice Hall, 1988. 439440 Bibliography
50. Daihatsu Motor Company, "Method of Control of NC Machine Tools", US patent 4445182, 1984.
51. Danielsson, P. E. "Incremental Curve Generation", IEEE

Transactions on Computers, Vol. C-19, No. 9, pp. 783–793, 1970.

52. Deitel, H. M. "An Introduction to Operating Systems", Addison Wesley, Boston, MA., 1984.
53. Deshayes, L., Donmez, A., Welsch, L., Ivester, R. "Robust Optimization for Smart Machining
54. Systems: An Enabler for Agile Manufacturing", Proceedings ASME International Mechanical Engineering Congress and Exposition, November 2005.
55. Fanuc, "Automatic Machining Process Determination Method in an Automatic Programming System", US patent 4723203, 1988.
56. Fanuc, "Part Profile Input Method", US patent 5089950, 1992.
57. Fanuc, "Acceleration/Deceleration Control Method for A Numerical Control Device", US patent 5218281, 1993.
58. Fanuc, "Graphic mechanism for checking interferences", Patent applied 94-700698, 1994.
59. Fanuc, "Conversational automatic programming system", Patent applied 94-702618, 1994.
60. Fanuc, "Numerical controller capable of estimating finishing time of machining operation", Patent applied 94-704020, 1994.
61. Fanuc, "Feedforward control method for a servomotor", US patent 5448145, 1995.
62. Fanuc, "Free-Form Curve Interpolation Method and Apparatus", US patent 5815401, 1998.
63. Fanuc, "Numerical control system", Patent applied 88-700333, 1988.
64. Fanuc, "Free Curve Interpolation Apparatus and Interpolation Method", US patent 5936864, 1999.
65. Final Report of ESPRIT Projects 6379 & 9115, "Open system architecture for controls within
66. automation systems: OSACA & Final report", ESPRIT, April 1996.
67. Fujita, S., Yoshida, T., "OSEC: Open System Environment for Controller", Proceedings Seventh International Machine Tool Engineering Conference, pp. 234–244, 1996.
68. GE Fanuc Automation North America, Inc., "Description Manual,

Series 21/210-Model B", 1996.

69. GMPTG, "Open modular architecture controls at GM POWERTRAIN - Technology and
70. Implementation: version 1.0", White paper downloaded from http://www.arcweb.com/omac, May 1996.
71. H¨ agglund, T., Astr¨ ° om, K. J. "Industrial Adaptive Controller Based on Frequency Response
72. Techniques", International Federation of Automatic Control, pp. 599–609, 1991.
73. Hardwick, M. "Justifying the STEP-NC savings", Fourth MDICM IRB Meeting, June 2001.
74. Heusinger, S., Rosso-Jr, R., Klemm, P., Newman, S., Rahimifard, S. "Integrating the CAx
75. process chain for STEP-compliant NC manufacturing of asymmetric parts", International
76. Journal of Computer Integrated Manufacturing, Vol. 19, No. 6, pp. 533–545, 2006.
77. Hughes Aircraft Company, "Robot Axis Controller Employing Feedback and Open Loop (feed forward) Control", US patent 4912753, 1990.
78. Hurco Companies, Inc., "CNC Control System", US patent 5453933, 1995.
79. Hyundai Motor Company, "Users' manual for HiTROL-M100", 2000.
80. Hyundai Motor Company, "Programming manual for HiTROL M-100", 2000.
81. Hyundai Motor Company, "Operating manual for HiTROL M-100", 2000.
82. Hyundai Precision Co. Ltd, "Implementation method for CNC controller using single CPU", Korea Patent applied 99-79125, 1999. Bibliography 441
83. Hyundai Precision Co. Ltd, "CNC controller having real time measurement capability", Korea Patent applied 99-79804, 1999.
84. Hyundai Precision Co. Ltd, "Turning CNC with tool path generation

method", Korea Patent applied 99-74236, 1999.

85. ICOM, "Method and Apparatus for Creating Custom Displays for Monitoring Ladder Logic Programs", US patent 4991076, 1991.
86. Im, S., Kang, S., Choi, J. "Machining cycle for rough cut and finish cut of turning operation with collision-avoidance capability", Proceedings of Conference of KSPE, pp. 1050–1053, 1997.
87. ISO/FDIS 14649-1, "Industrial automation systems and integration – Physical device control Data model for computerized numerical controllers – Part 1: Overview and fundamental principles", ISO/TC 184/SC 1/WG 7, 2002.
88. ISO/FDIS 14649-10, "Industrial automation systems and integration – Physical device control – Data model for computerized numerical controllers – Part 10: General process data", ISO/TC 184/SC 1/WG 7, 2004.
89. ISO/FDIS 14649-11, "Industrial automation systems and integration – Physical device control – Data model for computerized numerical controllers – Part 11: Process data for milling", ISO/TC 184/SC 1/WG 7, 2004.
90. ISO/FDIS 14649-12, "Industrial automation systems and integration – Physical device control – Data model for computerized numerical controllers – Part 12: Process data for turning", ISO/TC 184/SC 1/WG 7, 2005.
91. ISO/FDIS 14649-111, "Industrial automation systems and integration – Physical device control – Data model for computerized numerical controllers – Part 111: Tools for milling machines", ISO/TC 184/SC 1/WG 7, 2004.
92. ISO/IS 14649-121, "Industrial automation systems and integration – Physical device control Data model for computerized numerical controllers – Part 121: Tools for turning machines", ISO/TC 184/SC 1/WG 7, 2005.
93. Kang, C. "Machine tool technology: present and prospect", Journal of Korea Society of Precision Engineering, Vol. 13, No. 3, pp. 9–25, 1996.
94. Kang, J., Suh, S., "Machinability and set-up orientation for 5-axis numerically controlled machining of free surfaces", International Journal of Advanced Manufacturing Technology, Vol. 13, No. 5, pp.

311–325, 1997.

95. Kang, S., Lee, J., Choi, J. "Shop floor programming system for turning operations", Proceedings Conference of KSPE, pp. 707–712, 1995.
96. Kang, S. "NURBS interpolator for open architecture CNC control system", Proceedings
97. Eleventh Conference of KACC, pp. 656–659, October 1996.
98. Kim, D., Song, J., Kim, S. "Software-based acceleration and deceleration method for CNC machine and industrial robot", Journal of Korea Electrical Engineering, pp. 562–572, 1992.
99. Kim, H. J., Kim, G. T., Choi, B. W. "Architecture Design for Open Intelligent Manufacturing System", Proceedings Fourth IFAC Workshop, pp. 335–338, 1997.
100. Kim, H. "Characteristic analysis for system development", Proceedings Conference of KACC, pp. 22–27, 1998.
101. Kim, J., Im, Y., Chung, S. "Real time control system for multi-motor manipulation", Proceedings Eleventh Conference of KACC, pp. 1048–1051, 1996.
102. Kim, S. "Asymptotic PID control based on fuzzy inference and modified Ziegler–Nichols method", Journal of KSPE, Vol. 13, No. 5, pp. 74–83, 1996.
103. Kim, S. "Small motor control", Sungandang Publishing Co., Seoul, 2000.
104. Koren, Y. "Interpolator for a Computer Numerical Control System", IEEE Transactions on Computers, Vol. C-25, No. 1, pp. 32–37, 1976.
105. Koren, Y. "Cross coupled biaxial computer control for manufacturing system", ASME Transactions, Journal of Dynamic Systems, Measurement and Control, Vol. 102, No. 4, pp. 265– 272, December 1980.
106. Koren, Y., Masory, O. "Reference-pulse circular interpolators for CNC systems", Journal of Engineering for Industry, Vol. 103, pp. 131–136, 1981.442 Bibliography
107. Koren, Y. "Computer Control of Manufacturing System", McGraw Hill, New York, NY.,1983.
108. Koren, Y., Lo, C. "Variable gain cross coupling controller for

contouring", CIRP Annals, Vol. 104, pp. 371–374, August 1991.

109. Koren, Y., Lo, C. "Advanced Controllers for Feed Drives", CIRP Annals, Vol. 41, pp. 689-697, 1992.

110. Koren, Y., Pasek, Z. J., Ulsoy, A. G., Benchetrit, U. "Real-Time Open Control Architecture

111. and System Performance", CIRP Annals, Vol. 45/1, pp. 377–380, 1996.

112. Koren, Y. "Control of machine tools", Journal of Manufacturing Science and Engineering, Vol. 119, pp. 749–755, November 1997.

113. Kuo, B. C. "Automatic Control Systems", Fourth Edition, Prentice-Hall, Upper Saddle River, NJ., 1981.

114. Lavallee, R. "Soft Logic's New Challenge: Distributed Machine Control", Control Engineering, pp. 51–58, August 1996.

115. Lee, B. "NC readings", Sungandang Publishing Co., Seoul, 1995.

116. Lee, J., Park, H., Huh, W. "Tracking control of servo system with two degrees-of-freedom", Proceedings Eleventh Conference of KACC, pp. 844-847, 1996.

117. Lewis, R.W. "Programming Industrial Control Systems using IEC 1131-3", IEC, 1995.

118. LG Industrial Electronics Ltd., "Method for generating acceleration and deceleration for CNC machine tools", Patent applied 96-15124, 1996.

119. Lo, C. C. "Feedback Interpolators for CNC Machine Tools", Journal of Manufacturing Science and Engineering, Vol. 119, pp. 587–592, 1997.

120. Lynch, M. "Computer Numerical Control, Advanced Techniques", McGraw Hill, New York, NY., 1993.

121. Lyu J., Kang S., Chon, Y. "Minimum time pocket machining cycle for conversational programming system", Proceedings Conference of KACC, pp. 848–851, 1996.

122. Mazak, "Programming Manual for Mazatrol T32-2", 1990.

123. MDSI, "Open CNC Demo V5.1.1", www.misi2.com, 2000.

124. Mitsubushi Denki, "Numerical control machining method", Patent applied 84-6114, 1984.

125. Mitsubishi Denki, "Numerically Controlled Machining Method using Primary and Compensating cutters", US patent 4713747, 1987.

126. Mitsubishi Denki, "Machining data editing method for CNC controller", Patent applied 91-1508, 1991. Na, I., Choi, J., Chang, T., Choi, B., Song, O. "Contouring error analysis for PID control of

127. machining center", Journal of ICASE, pp. 32–39, 1997.

128. NAF Controls AB, "Method and an Apparatus in Tuning a PID Regulator", US patent 4549123, 1985.

129. National Instruments Corp., "System and Method for Closed Loop Autotuning of PID Controllers", US patent 6081751, June 2000.

130. NRL-SNT, www.step-nc.com, 2008.

131. Newman, S., Ali, L., Brail, A., Brecher, C., Klemm, P., Liu, R., Nassehi, A., Nguyen, V.,

132. Proctor, F., Rosso-Jr, R., Stroud, I., Suh, S., Vitr, M., Wang, L., Xu, X. "The Evolution of CNC

133. Technology from Automated Manufacture to Global Interoperable Manufacturing", Second

134. International Conference on Changeable, Agile, Reconfigurable and Virtual Production, July 2007.

135. Newman, S., Nassehi, A., Kumar, T., Vichare, P. "Universal Manufacturing Platform for

136. Global CNC Machining", International Workshop on Ubiquitous Manufacturing, February 2008. Ohyama, M., Tamai, S. "AC servo system: theory and design in practice", Jonghap Publishing Co., Seoul, 1990.

137. Oki Electric, "System for Interpolating an ARC for a Numerical Control System", US patent 4243924, 1981.

138. Olsson, G., Piani, G. "Computer Systems for Automation and Control", Prentice Hall, 1992. www.arcweb.com/omac, 2000. Bibliography 443

139. OMG, "CORBA 3.0 New Components Chapters OMG TC Document ptc/99-10-04", October 1999.

140. OMG, "The Common Object Request Broker: Architecture and Specification Revision 2.4", October 2000.

141. OSE consortium, "Development of OSEC (Open System Environment for Controller)",

142. OSEC Project technical report, October 1998.

143. Park, C. "Cycle machining method for numerical control", Patent applied 96-5564, 1996

144. Park, S., Jung, S. "Priority determination method under real time CORBA environment",

145. Journal of Korea Electrical Engineering, Vol. 38, No. 4, pp. 59–71, 2001.

146. Piegl , L., Tiller, W. " The NURBS Book", Second Edition, Springer Verlag, London, 1995.

147. Poo, A. N., Bolinger, J. G., Younkin, G.W. "Dynamic errors in type I contouring systems",

148. IEEE Transaction on Industrial Application, Vol. T.8, No. 4, pp. 477–484, 1972.

149. Pritschow, G., Philipp, W. "Research on the Efficiency of Feedforward Controllers in M Direct Drives", CIRP Annals, Vol. 41/1, pp. 411–415, 1992.

150. Pritschow, G., Daniel, G., Junghans, G., Sperling, W. "Open System Controllers-A Challenge for the Future of the Machine Tool Industry", CIRP Annals, Vol. 42/1, pp. 449–452, 1993.

151. Richard, J., Nguyen, V., Stroud, I. "Standardization of the Manufacturing Process: IMS/EU

152. STEP-NC project about Wire EDM Process", Intelligent Manufacturing System Conference, May 2004.

153. Rober S. J., Shin, Y. C. "Modeling and Control of CNC Machine using A PC-based Open

154. Architecture Controller", Mechatronics, Vol. 5, No. 4, pp. 401–420, 1995.

155. Samsung electronic, "Acceleration and deceleration method for servo motor", Patent applied 96-24775, 1996.

156. Samsung electronic, "Tool path generation method for turning operations", Patent applied 98-40717, 1998.

157. Schmitz, D., Khosla, P. "CHIMERA: A Real-Time Programming

Environment for Manipulator Control", Proceedings IEEE Conference of Robotics Automation, pp. 846–852, 1989.

158. Schofield, S., Wright, P. "Open Architecture Controllers for Machine Tools, Part 1: Design

159. Principles", Journal of Manufacturing Science and Engineering, Vol. 120, pp. 417–424, 1998.

160. Servo Motor, www.tao.co.kr, 2001.

161. Shin, D., Chung, S. "Design of digital filter for acceleration and deceleration control of servo motor", Journal of KSPE, Vol. 14, No. 9, pp.52–60, 1997.

162. Shin, S., Suh, S., Stroud, I. "Reincarnation of G-code based part programs into STEP-NC for turning applications", Computer Aided Design, Vol. 39, pp.1–16, 2007.

163. Siemens, "Method for limiting the rate-of-change of acceleration in numerical driving systems", US patent 5073748, 1991.

164. Siemens, "Sinumerik 840D OEM package NCK: User's Manual", Preliminary Edition, March 1995.

165. Silberschatz, A., Peterson, J., Galvin, P. "Operating System Concepts", Third Edition, Addison Wesley, Boston, MA., 1991.

166. Srinivasan, K., Kulkarni, P. "Cross coupled control of biaxial feed drive servomechanism",

167. Transactions of ASME, Journal of Dynamic Systems, Measurement and Control, Vol. 112, No. 2, pp. 225–232, June 1990.

168. Srinivasan, K., Tsao, T. C. "Machine tool feed drives and their control: A survey of the state of the art", Journal of Manufacturing Science and Engineering, Vol. 119, pp.743–748, November 1997.

169. STEP tools, www.steptools.com, 1999.

170. Stewart, D., Volpe, R., Khosla, P. "Design of Dynamically Reconfigurable Real-Time Software using Port-Based Objects", IEEE Transaction on Software Engineering, Vol. 23, No. 12, pp. 759–776, 1997.

171. Stroud, I., Xirouchakis, P. "Strategy features for communicating aesthetic shapes for manufacturing", International Journal of Computer Integrated Manufacturing, Vol. 19, No. 6, pp. 639–649, 2006.

172. Suh, S., Lee, K. "A prototype CAM system for 4-axis NC machining of rotational-freesurfaces", Journal of Manufacturing Systems, Vol. 10, No. 4, pp. 322–331, August 1991.444 Bibliography

173. Suh, S., Cho, J., Hascoet, J. "Incorporation of tool deflection in tool path computation: simulation and analysis", Journal of Manufacturing Systems, Vol. 15, No. 3, 1996.

174. Suh, S., Lee, S., Lee, J. "Compensating probe radius in free surface modeling with CMM: simulation and experiment", International Journal of Production Research, Vol. 34, No. 2, pp. 507–523, 1996.

175. Suh, S., Lee, J. "Interference-free tool-path planning for flank milling of twisted ruled surfaces", International Journal of Advanced Manufacturing Technology, Vol. 14, No. 11, pp. 795–805, 1998.

176. Suh, S., Lee, J., Kim, S., "Multiaxis machining with additional-axis NC system: theory and development", International Journal of Advanced Manufacturing Technology, Vol. 14, No. 12, pp. 865–875, 1998.

177. Suh, S. "Numerical Control and Integrated Manufacturing Systems", Second Edition, POSTECH Press, September 1999.

178. Suh, S. "STEP-NC Technology: Present and prospect", Journal of KSPE, Vol. 17, No. 5, pp. 8–14, 2000.

179. Suh, S., Jih, W., Hong, H., Chung, D., "Sculptured surface machining of spiral bevel gears with CNC milling", International Journal of Machine Tools & Manufacture, Vol. 41, pp. 833–850, May 2001.

180. Suh, S. and Cheon, S. "A Framework for an Intelligent CNC and Data Model", International Journal of Advanced Manufacturing Technology, Vol. 19, pp. 727–735, 2002.

181. Suh, S., Lee, B., Chung, D., Cheon, S. "Architecture and implementation of a shop-floor programming system for STEP-compliant CNC", Computer Aided Design, Vol. 35, pp. 1069– 1083, 2003.

182. Suh, S., Lee, B. "STEP-Manufacturing Roadmap", Proceedings Korea CAD/CAM Conference, January 2004.

183. Suh, S., Chung, D., Lee, B., Shin, S., Choi, I., Kim, K. "STEP-compliant CNC system for turning: Data model, architecture, and implementation", Computer Aided Design, Vol. 38, pp. 677–688,

2006.

184. Suh, S., Stroud, I. "A new model for machine data transfer", ISO Focus, pp. 24–26, 2007.
185. Sung, W. "A study on surface machining with CNC machining center", MS Thesis, Seoul National University, 1998.
186. Szafarczyk, M., Klein, B., Szala, W. "Extension of Typical CNC Systems by External Controllers", CIRP Annals, Vol. 38/1, pp. 351–354, 1989.
187. Szyperski, C., "Component Software: Beyond Object-Oriented Programming", Addison Wesley, Boston, MA., 1999.
188. Teltz, R., Elbestawi, M. "Design Basis and Implementation of an Open Architecture Machine
189. Tool Controller", Transactions of NAMRI/SME, Vol. XXV, pp. 299–304, 1997.
190. Timmerman, M., et al., "Windows NT real-time extensions better of worse?", Real-Time Magazine 98-3, pp. 11–19, 1998.
191. Timmerman, M., et al., "Is Windows CE 2.0 a real threat to the RTOS World?", Real-Time Magazine 98-3, pp. 20–24, 1998.
192. Tomizuka, M. "Zero phase error tracking algorithm for digital control", ASME Transactions,
193. Journal of Dynamic Systems, Measurement and Control, Vol. 109, pp. 65–68, 1987.
194. Toshiba, "Numerical Control Unit", US patent 5994863, 1999.
195. University of Utah Research, "Method and system for spline interpolation, and their use in CNC", US patent 5726896, 1998.
196. U.S. Philips Corp., "Machining apparatus wherein ARC length along a tool path is determined in relation to a parameter which is a monotonic function of time", US patent 5321623, 1994. VenturCom, "RTX 4.1 manual", 1997.
197. Weck, M., Bibring, H. "Handbook of Machine Tools, Vol. 3: Automation and Control", John Wiley & Sons, Hoboken, NJ., 1984.
198. Wosnik, M., Kramer, T., Selig, A., Klemm, P. "Enabling feedback of process data by use of
199. STEP-NC", International Journal of Computer Integrated

Manufacturing, Vol. 19, No. 6, pp. 559–569, 2006.Bibliography 445

200. Xu, X., Wang, H., Mao, J., Newman, S., Kramer, T., Proctor, F., Michaloski, J. "STEPcompliant NC research: the search for intelligent CAD/CAPP/CAM/CNC integration", International Journal of Production Research, Vol. 43, No. 17, pp. 3703–3743, 2005.
201. Xu, X., Proctor, F., Klemm, P., Suh, S. "STEP-compliant process planning and manufacturing", International Journal of Computer Integrated Manufacturing, Vol. 19, No. 6, pp. 491– 494, September 2006.
202. Yamazaki, K. "Open Architecture CNC Controller in the USA", Proceedings Seventh International Machine Tool Engineering Conference, pp. 204–218, 1996.
203. Yang, D., Kong, T. "Parametric interpolator versus linear interpolator for precision CNC machining", Computer Aided Design, Vol. 26, No. 3, pp. 225–233, 1994.
204. Yeh, S., Hus, P. "The speed-controlled interpolator for machining parametric curve", Computer Aided Design, Vol. 31, No. 5, pp. 349–357, 1999.
205. Zahavi, R. "Enterprise Application Integration with CORBA: Component and Web-Based Solutions", John Wiley & Sons, 2000.
206. Ziegler, J. G., Nichols, N. B. "Optimum Settings for Automatic Controllers", ASME Transactions, pp. 759–768, 1942.
207. Ziegler, J. G., Nichols, N. B., "Process Lags in Automatic-Control Circuits", ASME Transactions, pp. 433–444, 1943.
208. Eric M. Taleff. St, P.A.F.P.E.K., 2004. Advances in superplasticity and superplastic forming: proceedings of a symposium : held during the 2004 TMS Annual Meeting, Charlotte, North Carolina, USA, March 14-18, 2004, Volume 1. Crafts & Hobbies.
209. Giuliano, G., 2011. Superplastic Forming of Advanced Metallic Materials: Methods and Applications. Elsevier Science.
210. GolamKibria,B.B.J.P.D.,ed.,2017.Non-traditionalMicromachining Processes: Fundamentals and Applications. Springer.
211. H. R. Parsaei, W.G.S., n.d. Economic and Financial Justification of Advanced Manufacturing Technologies. Elsevier.

212. Hans J. Scheel, P.C., ed., n.d. Crystal Growth Technology: From Fundamentals and Simulation to Large-scale production. Wiley.
213. Hyeon Il Park, D.K.J.L.J.H.K.M.-G.L.Y.L.J.H.S., 2014. Experimental Study on Electromagnetic Forming of High Strength Steel Sheets with Different Dimensions of Aluminum Driver Plate. AbeBooks.
214. Jha, N.K., n.d. Handbook of Flexible Manufacturing Systems. Academic Press..
215. Jr, F.C.C., 2011. Manufacturing Technology for Aerospace Structural Materials. Elsevier.
216. K Case, S.T.N., ed., n.d. Advances In Manufacturing Technology VIII: Proceedings Of The 10th National conference on manufacturing research. K Case, S T Newman.
217. K. G. Swift, J.D.B., 2013. Manufacturing Process Selection Handbook. Butterworth-Heinemann.
218. Koç, M., ed., 2008. Hydroforming for Advanced Manufacturing. Elsevier.
219. Kumar, H.S.L.K.F.A.S., n.d. sciencedirect. [Online] Available at: http://www.sciencedirect.com/science/article/pii/S0890695502000238 [Accessed 5 July 2017].
220. MBA, d.A.M., 2013. Green Manufacturing: (Paradigm Shift to Sustainable Capitalism). AA Global Sourcing Ltd.
221. MIMechE, T.S.C., n.d. researchgat. [Online] Available at: https://www.researchgate.net/publication/303841938_ELID_Grinding_and_EDM_for_Finish_Machining.
222. Norrish, J., n.d. Advanced Welding Processes. Elsevier. nptel.ac.in, n.d. npte. [Online] Available at: http://nptel.ac.in/courses/112107077/module2/lecture3/lecture3.pdf [Accessed 5 July 2017].
223. Prasad, B., n.d. Concurrent Engineering: Contemporary issues and modern design tools. Prentice Hall PTR.
224. R.L. Timings, S.P.W., 2009. Manufacturing Technology Volume Ii, 2/E. Pearson Education India.
225. Raja, V.F.K.J., ed., 2008. Reverse Engineering An Industrial Perspective. Springer.
226. Taylan Altan, A.E.T., 2012. Sheet Metal Forming: Processes and

Applications. ASM International.

227. Tianran Wang, Y.Z.H.Y.F.-Y.W., ed., 2012. Advanced Manufacturing Technology in China: A Roadmap to 2050. Springer Science & Business Media..

228. Waldemar Karwowski, G.S., n.d. Organization and Management of Advanced Manufacturing. Wiley.

Index

Symbols

A

B

C

E

V

W